Practical
Weather Forecasting

Practical
Weather Forecasting

Frank Mitchell-Christie

BARRON'S
Woodbury, N.Y.

First U.S. Edition 1978 by Barron's Educational Series, Inc.

First published in Great Britain in 1977 by
William Luscombe Publisher Limited
The Mitchell Beazley Group
Artists House
14–15 Manette Street
London W1V 5LB

Address all inquiries to:
Barron's Educational Series, Inc.
113 Crossways Park Drive
Woodbury, New York 11797

ISBN 0-8120-5210-2

Library of Congress Catalog Number 77-84142

Designed by Arka Graphics

PRINTED IN GREAT BRITAIN

Contents

Chapter 1 What do we mean by 'Weather'?

RELIABLE WEATHER FORECASTING is dependent on trained observers studying and reporting on all the various weather factors, every hour on the hour. Such observers man a vast network of reporting stations spread all over the world, and linked by an international teleprinter system. A typical weather station site has to be well-exposed to the wind, whatever its direction, and be equipped with the following instruments:

Anemometer to give wind speed and direction. It is linked electrically to an *Anemograph* which records these fluctuating values.

Stevenson screen: a white louvred box permitting air to flow freely through it but protecting the thermometer inside from the direct rays of the sun.

Thermometers used are *dry-bulb*, to give the actual air temperature; *wet-bulb* to give values of humidity and dew-point (see Glossary, page 15.); *maximum/minimum* to record highest temperature by day, and lowest at night; *ground* and *soil* thermometers to record ground temperatures. *Barometers* measure the pressure in the atmosphere.

All other weather information is collated by trained observers, using specifications and figure-codes internationally agreed, recorded, then fed into the international teleprinter network. National weather reports and forecasts are distilled from all this information.

In the chapters that follow these all-important weather factors are described and explained, showing how they affect us all, in whatever part of the world we live.

Defining the 'weather'

'Weather' has many aspects, each of differing importance to different people at different times, in different places, or for differing activities. It is therefore necessary to define exactly what we mean by the word 'weather', and to give precise meanings to the words used to describe the 'weather' in all its aspects.

The word 'weather' can be used in two distinctly different ways. Firstly, it can be used to mean all the aspects of atmospheric state and movement which can be seen or experienced, and which affect human activity. For example, the study of the weather must include wind, clouds, rainfall, temperature and, at sea, waves. All weather forecasts must give full details of all these aspects and must leave the reader or listener to select and give attention to whichever aspect of the total picture is of most importance to him and to his activities.

Secondly, the word 'weather' can be used with a more limited meaning, a meaning restricted to aspects to which the human body is always sensitive – namely, whether it will be sunny or dull, hot or cold, or whether it will rain or snow. This is the more common usage because these aspects have a personal impact on everyone. Indeed in countries such as Britain where rapid and marked changeability is the rule rather than the exception, these aspects of the weather often become the major topic of conversation!

Thus 'weather' in its total sense is made up of a number of factors of which 'weather' in its more restricted sense is just one. This becomes clear when all the factors are listed. Each of these factors must be understood if the 'weather' is ever to make sense.

The weather factors

Factor 1: WIND Where the moving air has come from, how long it has taken and what strength it has. The wind, of course brings:

Factor 2: WEATHER The state of the sky; the rain or snow which falls from the clouds in the sky; the state of the ground; the temperature and humidity – which in turn, depend on:

Factor 3: CLOUD The type, the amount, the height, the vertical extent and the rate of development or decay.

Factor 4: VISIBILITY Poor visibility curtails some activities, makes others very dangerous, and fog makes some impossible.

Factor 5: STATE OF THE SEA The waves created by the wind. This factor, as it were, completes the cycle for weather conditions at sea.

These weather factors are inter-related both as regards cause and effect. The study of the weather is therefore the study of these factors and of their relationships.

The factors are listed above in their logical order. The dominant factor, the Wind, comes first. The second is the Weather brought by the wind. The third is Cloud because the type, height and amount of cloud all greatly influence surface temperatures and Relative Humidity which in their turn affect the fourth factor, Visibility. The fifth factor, the state of the sea, depends on the first – the Wind.

Whatever aspect of the weather may be of special importance to you, always consider each of the above factors, and in the order indicated. Doing so will enable their inter-relationship to be fully appreciated – and errors of interpretation guarded against.

Typical weather station, showing how the various weather-recording instruments are arranged for maximum efficiency.

Beaufort wind scale – on land

Force	m.p.h.		
0	1	Light	Smoke rises vertically.
1	1–3	Light	Direction of wind shown by smoke drift. But not by ordinary wind vanes.
2	4–7	Light	Wind felt on face; leaves rustle; ordinary wind vane moves.
3	8–12	Gentle	Leaves and small twigs in constant motion wind extends light flags.
4	13–18	Moderate	Raises dust and loose paper; small branches move.
5	19–24	Fresh	Small trees in leaf begin to sway; small crested waves on inland waters.
6	25–31	Strong	Large branches in motion; telephone wires whistle; umbrellas difficult.
7	32–38	Strong	Whole trees in motion; inconvenience in walking against the wind.
8	39–46	Gale	Twigs break off trees; progress generally impeded.
9	47–54	Gale	Slight structural damages; tiles or slates go.
10	55–63	Storm	Trees uprooted; roofs damaged. Other structural damage.
11	64–72	Violent Storm	Widespread damage.
12	Over 72	Hurricane	Damage to 'Disaster' scale.

Beaufort wind scale at sea – summary

Force	Knots	Called	Sea Conditions	Probable Wave Height in Feet Mean	Max.
0	< 1	Calm	Sea like a mirror.		
1	1–3	Light air	Ripples with the appearance of scales formed but without foam crests.	$\frac{1}{4}$	—
2	4–6	Light breeze	Small wavelets, still short but more pronounced, crests have a glassy appearance and do not break.	$\frac{1}{2}$	—
3	7–10	Gentle breeze	Large wavelets. Crests begin to break. Foam of glassy appearance.	2	3
4	11–16	Moderate breeze	Small waves, becoming longer; fairly frequent white horses.	$3\frac{1}{2}$	6
5	17–21	Fresh breeze	Moderate waves. taking a more pronounced long form; many white horses are formed. Chance of some spray.	6	10
6	22–27	Strong breeze	Large waves begin to form; the white foam crests are more extensive everywhere. Probably some spray.	$13\frac{1}{2}$	18
7	28–33	Near gale	Sea heaps up with white foam from breaking waves.	18	24
8	34–40	Gale	Moderately high waves of greater length; much foam.	23	30
9	41–47	Strong gale	High waves. Dense streaks of foam along the direction of the wind.	25	33
10	48–55	Storm	Very high waves with long overhanging crests. Visibility affected.	29	40
11	56–65	Violent storm	Exceptionally high waves. Visibility seriously affected.	37	50
12	67–	Hurricane	The air is filled with foam and spray. Visibility very seriously affected.	45	60

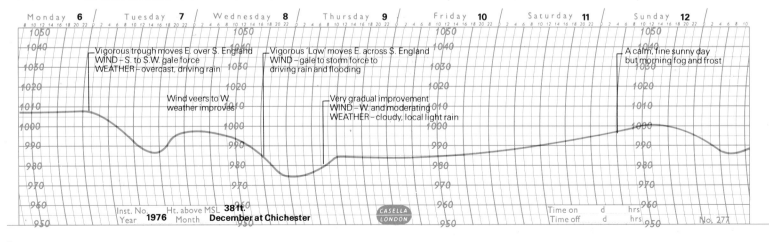

Monday 6 | Tuesday 7 | Wednesday 8 | Thursday 9 | Friday 10 | Saturday 11 | Sunday 12

Vigorous trough moves E. over S. England
WIND – S. to S.W. gale force
WEATHER – overcast, driving rain

Vigorous 'Low' moves E. across S. England
WIND – gale to storm force to driving rain and flooding

A calm, fine sunny day but morning fog and frost

Wind veers to W. weather improves

Very gradual improvement
WIND – W. and moderating
WEATHER – cloudy, local light rain

Inst. No.   Ht. above MSL 38 ft.
Year 1976 Month December at Chichester

CASELLA LONDON

Time on   d   hrs
Time off   d   hrs

No. 277

Fig. 1 (bottom page 8). Barograph trace covering seven days in early December, 1976 and annotated with notes on the weather that week.

Fig. 2 (right). Kew pattern mercury barometer, with (A) the 'Gold' slide used to make corrections and adjustments for differences in latitude and height etc.

Fig. 3 (below). Pressure and vernier scales on the Kew pattern barometer.

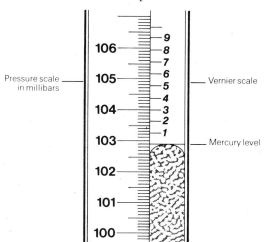

Pressure scale in millibars

Vernier scale

Mercury level

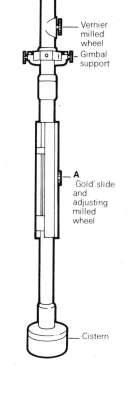

Pressure scale in millibars — Vernier scale

Vernier milled wheel — Gimbal support

A Gold' slide and adjusting milled wheel

Cistern

Fig. 4 (right). Exploded and simplified view of an aneroid barometer. High pressure compresses the vacuum box bottom left against the spring. This pulls down on the wire which then moves the pointer round towards 'high' on the face of the barometer. When pressure is low, the reverse happens.

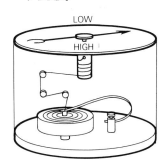

LOW
HIGH

Wind and pressure

Wind speed is usually expressed in:
1. The Beaufort Scale, as a force number.
2. Knots (nautical miles per hour). 60 nautical miles equal 1° of latitude. e.g. A wind force 8, will be blowing at a speed of between 34–40 knots (27–36 mph). Wind speed is determined by pressure difference called the *Gradient*.

Wind direction is usually expressed:
1. To the nearest 10° of the 0°–360° scale.
2. By compass points (32 in all). *Note:* N by E is between N and NNE; it is *not* NE.

One would expect wind to blow directly from High to Low pressure, but due to the spin of the earth, it is in fact deflected along the *isobars* – which are lines on the weather chart, along which pressure is equal. The spacing of the isobars determines the wind speed.

Gustiness occurs where there is a variation in both wind speed and direction.

Veering is a clockwise change in direction of the wind, e.g. from S to SW. In the Northern hemisphere this is usually sudden and often violent.

Backing is a counterclockwise change of direction of wind, usually gently in the Northern Hemisphere, but sudden and violent in the Southern.

Freshening indicates an increase in wind speed.

Moderating indicates a decrease in wind speed.

Pressure is usually expressed in:
1. Inches of Mercury or lbs/sq. in.
2. *Millibars*, where 1,000 mbs = 1 *Bar* = 1 *Atmosphere.*

The 'Weather'

As a weather factor this heading covers:
(a) The state of the sky. (b) Precipitation. (c) Fog, mist, haze. (d) Temperature and humidity. (e) State of the ground. (f) Local weather hazards.

(a) The state of the sky
Basically there are only four words to describe the weather – *fine, fair, cloudy* and *overcast* – and they are related to the amount of the sky covered by cloud.

Fine: the sky with 2/8, or less of cloud. Beaufort letter **b**, for *Blue Sky*.

By day there will be 75% or more of possible hours of sunshine. Temperatures, both thermometer and by human body assessment, will be high provided there is not too much wind. Some places will experience 100% sunshine from sunrise through to sunset. In general the quality of light will be good, shadows will be sharp and visibility will also be good.

By night there will be a maximum cooling from a surface unprotected by cloud cover, leading to low minimum temperature at, or just after, dawn which will then lead to the formation of dew, of ground mist, or of frost.

Both by day and by night navigation is at its easiest: all the stars are visible at night, and, near land, all the coastal navigation aids will be clearly recognisable at maximum distances.

Fair: 2/8 to 4/8 of cloud. Beaufort letters **bc**, a mixture of blue sky and some cloud.

By day, not quite so hot, but still about 50% sunshine.

By night, not quite so cold, and the navigator has about 50% of the stars to 'shoot'.

Cloudy: 4/8 to 6/8 of cloud. Beaufort letter **c**, for cloud.

By day, the sunshine hours are reduced to much less than half, the clouds reflecting the sun's heat back into space to give a cool, if not a cold, day.

By night the clouds reflect the heat radiating from the earth's surface back to the earth – on a mild night, a little dew, perhaps, but little likelihood of fog or of frost.

At sea, the navigator will have to watch for his stars but he should get a reasonably good fix of position. Nearer land the reflection of a lighthouse beam from the cloud may enable him to sight that light over the curvature of the earth's surface at well beyond its geographical range – a bonus for him.

Overcast: The sky almost completely covered by cloud. Beaufort letter **o** for overcast.

By day, all the sun's heat is reflected off the upper surface of the cloud, the surface is denied that heat and temperatures are unseasonably low.

By night, the cloud blanket traps in the radiation of heat from the surface and temperatures will be unseasonably high. To the ocean navigator the implications are more serious.

If the overcast cloud layer is thin he may still sight the sun or the moon through it so that by day he can get a good fix of position and a moderately good single position line from the moon at either evening or morning twilight.

But if the cloud is thick and the cloud-base low he will sight nothing, a circumstance which can continue for several days and so deny him the comfort of an accurate fix of position.

Furthermore very low cloud will cover coastal hills and headlands (as hill fog) and will render even the most familiar coast-line completely unrecognisable. This hazard is equally serious to mountaineers and to hikers. And they, unlike the modern navigator, do not enjoy the benefits of electronic aids!

(b) Precipitation
Precipitation is what falls out of the clouds – and what falls is water; either as rain, snow or the mixture we call *sleet*; or as *drizzle*; or as *hail*.

If the 'weather' (referring to the state of the sky), is described as *fine* or *fair* there cannot be precipitation, for this can only occur in *cloudy* or *overcast* conditions.

Showers, of *rain, hail* or *snow,* are localised precipitation from recognisably individual clouds or clusters of clouds, and these clouds will be of the Cumulus family. (See page 13.)

Extensive precipitation, over a large area and persisting for several hours comes from the Stratus family of clouds. Its form (drizzle, rain, sleet or snow) depends on a combination of temperature, the rate of formation, and height of the cloud layer from which it is falling.

Intensity of precipitation can be described as light, moderate or heavy; *duration* as intermittent or continuous.

Types of precipitation
DRIZZLE.   Very small droplets, forming and falling slowly.
RAIN.   Larger drops, heavy enough to fall quickly yet light enough to be slanted by the wind.
SNOW.   Complex ice-crystals created when minute rain-drops fall into and down through a lower layer of air at freezing point or colder. The crystals form slowly and, colliding, join together to form flakes.

Whether the snow melts into slush or lies will depend on whether the surface temperature is marginally above 32 °F (0 °C) or not. Snow lying to an even depth of 1 ft. (30 cm), is equivalent to 1 in. (2.5 cm) of rainfall.
SLEET.   An uncomfortable mixture of rain and snow.
HAIL.   Solid balls of ice varying in size, formed by a sequence of up and down journeys in the turbulence of towering Cumulus clouds, crossing and re-crossing the freezing-level and eventually falling too fast for re-evaporation to take place. Hail is always a localised 'shower', from one cloud.

(c) Fog, mist and haze as weather factors
Fog, mist and haze are surface phenomena which affect visibility. Consequently it is important they are included in weather reports and forecasts.

Haze is composed of dust particles, and in country areas rarely reduces visibility to less than three miles. However, near to and downwind of industrial areas, chemicals discharged into the atmosphere can lead to an increase in both the number and the size of air-borne particles. Industrial haze can then reduce visibility to much less than one mile (1.6 km.) and it can extend for several hundred miles/kilometres. Haze is usually associated with fine weather when thermal up-currents do not exist to spread the dust upwards and thin it out.

Mist and *fog* are in essence 'clouds on the ground' – minute water droplets breaking the paths of light rays and reducing visibility to the point where human activities are at best somewhat restricted and at worst reduced to a standstill or made very dangerous.

(d) Temperature and humidity
By their degree and value, these two important components of the weather contribute vitally towards all the other factors listed in this section.

(e) State of the ground

Many people – not only farmers – need to know how the weather is likely to affect the land, and what ground conditions can be expected.

Dew occurs when the surface temperature falls to the point where water-vapour condenses on the cold surface to form water drops. The most favourable conditions are fine clear nights with little or no wind, giving maximum radiation cooling and with still, cold air concentrated at or just above the surface. Heavy dew can provide sufficient moisture for some crops during long periods of rainless fine weather.

The Dew-point is the temperature at which humidity reaches 100% (i.e. the air can hold no more water-vapour), and condensation takes place. When this occurs at ground level *dew* forms; when it occurs throughout a thicker layer of surface air, *fog* forms; when it occurs higher in the sky, *cloud* forms.

Hoar frost occurs when the temperature of the radiating surface is below freezing point and instead of dew-drops, ice-crystals form.

Rime: When frost and fog occur together, the water drops of the fog can become super-cooled to freeze instantaneously if they come into contact with such objects as tree-branches, overhead wires, car windscreens, aircraft wings. A deposit of rough white ice accumulates to windward of all such exposed surfaces.

Glazed frost is a transparent, almost invisible, smooth coating of ice formed in much the same way as hoar frost and rime except that the process takes place quickly. It can occur when rain falls on very cold surfaces, or when the temperature of damp surfaces falls quickly to below freezing, e.g. on wet roads when damp patches become treacherous patches of ice.

Ground frost is forecast when a thermometer placed horizontally one inch above ground is expected to fall below minus 1°C (30°F) – when either hoar frost or rime will probably form, depending on local humidity values and wind speed.

The important aspect of this low ground temperature is that plant growth is likely to be seriously affected.

Air frost is forecast when the temperature between the ground and tree-top level (30–50 ft.; 10–15 m.) is expected to fall below zero. This condition occurs with a wind speed of Force 1 to 2 – just enough to disturb the coldest air from the ground and lift it in gentle swirls into the trees – to the detriment of fruit blossom.

Local weather hazards

Some weather phenomena are geographically very small in size but often violent and destructive.

These hazards are:

WHIRLWINDS OF TORNADOES.

WATERSPOUTS.

DUST STORMS

GALE FORCE WINDS or stronger.

SUDDEN CHANGES IN DIRECTION of strong winds.

STORM SURGES – abnormally high or low tides.

FLOOD WARNINGS related to abnormally high rainfall, especially if coinciding with high tides or high water-levels in rivers.

Cloud as a weather factor

The terms *fine*, *fair*, *cloudy*, *overcast* either used separately or in combination (e.g. cloudy with fair periods) describe the state of the sky in a way that is adequate for many human activities. However, additional information about the clouds in the sky is always useful and sometimes essential (e.g. for flying and mountaineering). Weather reports and forecasts therefore include:

1. The type or types of cloud present or forecast.
2. The height of the cloud-base and vertical development.
3. The amount of the sky covered by each type or layer of cloud, expressed in tenths.

Cloud Classification

Clouds are classified into ten main types (*genera*) according to height and 'family'.

Height division	Usual altitude		Genera names
	Feet (000's)	Metres (000's)	
HIGH (Cirro)	20–40	6–12	Cirrus
	20–40	6–12	Cirro-stratus
	20–40	6–12	Cirro-cumulus
MEDIUM (Alto)	6–20	2–6	Alto-stratus
	6–20	2–6	Nimbo-stratus
	6–20	2–6	Alto-cumulus
LOW	0–2	0–.6	Stratus
	1–5	.3–1.5	Cumulus
	1–4.5	.3–1.35	Strato-cumulus
	1–5	.3–1.5	Cumulo-nimbus

The two main families in the ten classified *genera* listed in the table are:

Stratus: the layered clouds which develop when warm moist Tropical air masses run up and over heavier cold Polar air to give frontal cloud and rainfall over a wide geographical area.

Cumulus: the towering, billowing convective cloud which develops in cold, unstable Polar air-streams, and gives localised shower precipitation.

From the latin word *Nimbus* (meaning *rain*), come the names of two other *genera*:

Nimbo-stratus: when the condensation rate becomes vigorous enough in a layer of alto-stratus cloud for rain to fall, that cloud, by definition, becomes nimbo-stratus. Heavy rain can cause its base to lower, when it would of course be re-classified as Low Cloud.

Cumulo-nimbus: Similarly, when the formation of hail stones in the up-down turbulence of towering cumulus clouds becomes rapid enough, a show of hail or rain will fall from that cloud. Cumulus clouds are ripe for thunder to occur when they extend upwards for more than 12,000 ft. (5,500 m.). Eventually, at very high altitude strong winds blow off their tops into the classical 'anvil' shape. This may persist as 'false cirrus' long after the main cloud has dissipated.

Strato-cumulus: this is in a class of its own, neither layered nor billowing; the air-stream neither wholly stable nor wholly unstable. Precipitation from this cloud is usually intermittent and always light.

Cirrus: are the 'Mare's tails' or wisps of ice-crystals perhaps presaging the approach of a storm, or perhaps the last remaining traces of the thunder-tops of the last storm. In their many varieties, they also form a class of their own.

Cloud base

Cloud base and *vertical extent* are primarily of importance in aviation but sailors should also possess information about them. A low layer of stratus cloud will pick up and reflect the beam of a powerful lighthouse, enabling it to be seen as a 'loom' and a bearing taken of its direction far beyond the normal range as limited by the curvature of the earth.

For example, the range of a lighthouse 225 ft. (70 m.) high is about 16 miles (25 km.). A well-defined cloud layer 900 ft. (275 m.) higher would create a 'loom' visible at over 30 miles (48 km.); a cloud base of 2,500 ft. (762 m.) would extend it to 50 miles (80 km.). These figures are dependent on good visibility beneath the cloud, but unless this phenomenon is understood sailors can sight lights at distances very much greater than might be thought possible, with misleading results.

Another hazard arises when cloud is low enough to obliterate the hills of a coast-line, thus making the most familiar land-fall completely unrecognisable.

Visibility as a weather factor

On land few activities are curtailed until visibility falls below 600 ft. (180 m.) due to fog. Some are however, and those engaged in them are entitled to accurate forecasts of maximum likely visibility.

At sea, visibility is of prime importance. If less than 1,000 metres (just over 1,000 yards) it is officially FOG, and all vessels, irrespective of electronic aids, are required by law to reduce speed and make noises indicating where they are and what they are doing.

Visibility is also an important factor in navigation at sea, since a clear horizon is needed as a base-line for the navigator's sextant. In coastal waters where he relies on visual bearings, he is virtually 'blind' when visibility is reduced to 1–2 miles (1,600–3,200 m.).

Weather factors at sea

The waves created by the wind blowing over the sea are in harmony with the wind strength:

Beaufort Force	Sea State
0	Smooth
1–3	Calm
4	Slight
5	Moderate
6–7	Rather Rough
8	Rough
9–12	Very Rough

Swell refers to waves existing from a wind which has died away, or created by winds blowing over a distant generating area. It is described by height as *low*, *medium* or *high* and by its wave-length as *short*, *moderate* or *long*.

Because swell waves attain speeds approaching those of the winds which generate them, they run faster than the windbelt itself moves. Hence they run ahead of storms, and are certain proof of very strong winds somewhere in the direction from which the swell is running – a warning that should not be ignored.

In this first chapter the weather factors have been defined and listed. Little explanation of cause and effect has yet been attempted. This comes next, in studying the reasons why the wind blows and how it blows, why clouds form and why rainfall occurs.

Glossary of terms

AEROGRAM, AEROLOGICAL DIAGRAM A graphical representation of pressure, temperature and humidity upwards through the atmosphere.

ANABATIC WIND Local wind which blows up a slope heated by sunshine.

ANTI-CYCLONE or HIGH PRESSURE An area where the pressure is high and the wind circulation is in the opposite direction to that of a cyclone (low pressure) i.e. clockwise in the northern hemisphere and counterclockwise in the southern.

BUYS BALLOT'S LAW This relates wind direction to pressure, stating that in the northern hemisphere an observer with his back to the wind has low pressure on his left i.e. winds around Lows are counterclockwise.

CLOUD Collection of water droplets or ice crystals in suspension in the air, resulting when the air is cooled to its dew point and condensation occurs.

COLD FRONT The frontier of Polar air as it surges into temperate latitudes, usually swinging in from the west behind depressions.

CONVECTIVE CLOUD, RAIN Heated air will rise and cool to its dew point to form cloud. If the process is rapid, water droplets will fall as rain.

CUMULUS The accumulating or billowy family of clouds.

CYCLONE or LOW PRESSURE Either an area of low pressure or the counterclockwise (in northern hemisphere) swirl of winds around that low pressure area, or both, according to context. In the tropics emphasis is on the strong winds generated and the local names reflect this i.e. *Typhoon* = big wind. In the temperate latitudes cyclones are usually called Lows or Depressions.

DEW POINT The temperature at which air becomes saturated (i.e. can hold no more water vapour), Relative Humidity is 100%, condensation takes place and latent heat is released.

DOLDRUMS Equatorial sea areas of calms or very variable winds, and of violent squally thunderstorms. Now called the Inter-tropical Front or I.T. Convergence Zone.

GRADIENT PRESSURE Just as closely-spaced contour lines on a map indicate a steep gradient on land, so closely-spaced isobars (q.v.) indicate a steep pressure gradient and resulting strong winds.

GULF STREAM The flow of warm tropical surface water out of the Gulf of Mexico through the Straits of Florida. It spreads north and east to affect the climate of Europe.

HIGH PRESSURE (SEE ANTI-CYCLONE)

HORSE LATITUDES 30°–40°N. or S. latitudes, where sub-tropical oceanic anti-cyclones (Highs) give long periods of calms or very light variable winds. The name arose because voyages under sail would often become prolonged, and livestock, especially horses, had to be jettisoned through lack of water etc.

HUMIDITY Dampness or water-vapour content of the air, most usually expressed in terms of *Relative Humidity* i.e. a percentage of the amount of water vapour which would totally saturate that portion of air at that temperature.

HURRICANE Tropical cyclone of the West Indies. Beaufort Force 10.

INVERSION It is normal for temperature to fall the higher one goes into the atmosphere. When this process is reversed and thermal uplift is consequently prevented, it is called 'inversion'.

ISOBARS Lines of equal pressure on the weather map. Their likeness to contours on a topographical map have resulted in such geographical terms as ridge, col and trough being adopted into meteorology.

KATABATIC Cold air flowing rapidly downhill, especially off high mountain ranges.

LOW PRESSURE (SEE CYCLONE)

NIMBO/NIMBUS Rain-bearing – hence *nimbo-stratus* and *cumulo-nimbus* cloud.

OCCLUSION This occurs when the Polar air behind a depression catches up with the returning Polar air ahead, lifting the warm Tropical air (shut-off) above the land surface. (See *Warm Sector*).

OROGRAPHIC CLOUD/RAIN Occurs when an air stream lifts over rising ground.

POLAR FRONT The 'frontier' between Polar and Tropical air between latitudes 40°–60°N. and S., along which temperate latitude depressions (cyclones) form and swirl rain-bearing cloud eastwards over the continents.

PRESSURE The modern unit of measurement is the Millibar, equal to 1,000 Dynes (C.G.S. units). This has replaced inches of mercury.

SQUALL A strong gusty local wind, usually associated with a cumulo-nimbus cloud. It rises suddenly, lasts only a few minutes, and as quickly dies away.

STRATUS The layered cloud family.

TROPICAL CYCLONES Small (usually less than 500 miles across), deep (pressure below 950 mbs.), very vigorous (wind speeds up to force 10 or even 12), swirls of hot, moist tropical air forming about 300 miles from the equator, at first moving slowly westwards between latitudes 20°–30° at about 10–15 knots, before re-curving away to the east. Local names are hurricane and typhoon.

WARM FRONT The frontier of Tropical air as it presses Pole-wards.

WARM SECTOR The sector of temperate latitude depressions which have warm Tropical air at the surface.

Chapter 2 The wind and clouds

WIND IS THE atmosphere in motion. Less than ten miles thick (compared with the 8,000 miles' diameter of the Earth), the atmosphere is a very thin envelope of air surrounding the Earth. It is in constant motion, moving over the face of the Earth in complex swirls. Some of these swirls are vast in size and remarkably constant both in space and time; others are smaller but are often the most violent.

In some parts of the world the direction and speed of the wind continue almost constant for days or weeks on end, and the weather accordingly remains likewise. In other places the wind's direction and speed are constantly changing, as swirls of varying size and intensity pass overhead; the weather is then equally changeable. Sometimes changes of direction are not only sudden but of great magnitude. Shifts of between 45 and 90 degrees, taking place in a matter of minutes, are commonplace in the swirls of the Temperate Latitudes called 'depressions', and even greater shifts of 180 degrees or more are not infrequent. It is these violent shifts of direction, rather than the strength of the winds involved (although they are usually associated with winds of high force), which cause the most damage.

What makes the wind

The atmosphere – in a constant state of movement – reacts continually to two forces which together determine which way it will move and at what speed. The first of these forces is *the weight of all the other air above it*. At the Earth's surface this is the total weight of the atmosphere above any given point – a weight which we measure with a *barometer* (*Baros* is Greek for weight), call *Pressure* and express in a variety of units: pounds per square inch, inches or centimetres of Mercury, or Millibars. This first force moves heavy air towards lighter airs; the wind blows from *High* to *Low* Pressure. The greater the pressure difference between High and Low – the *Pressure Gradient* – the stronger blows the wind, so that the Pressure Gradient determines the strength of the wind. This is called the *Gradient Wind*.

The second force acting on the atmosphere is that of *the rotation of the Earth*. The spinning surface of the Earth is moving at 900 knots (1,000 m.p.h.) at the Equator, at half that speed in Latitude 60′N or 60′S, reducing to nil at the Poles. This second force deflects the air as soon as it is in motion.

As soon as any air is in motion under the first force, the second force comes into operation. Called the *Coriolis* or *Geostrophic* (literally 'world-turning') force, its effect is to alter the direction of movement, to the *right* in the Northern Hemisphere and the *left* in the Southern. The magnitude of this force, at the present speed of rotation of the Earth, is such that the deflection is 90°. As a result the *direction* of the wind is not across the isobars directly from High Pressure to Low, but along them at right angles to the Gradient.

A second effect of this force, varying as it does with Latitude, is to decrease the speed of the Gradient Wind as one moves from the Equator towards the Poles.

Geostrophic wind

This is the name given to the air movement resulting from the action and reaction of these two forces. Its direction is always along the *isobars* (lines of equal pressure), and its speed is proportional to the Pressure Gradient, that is, to the spacing of the isobars.

Geostrophic wind scale
On all Meteorological Charts a Scale is given showing this basically important relationship between the spacing of the isobars and the actual Wind Speed over the Earth's surface. The Scale can be used both ways – from known isobaric spacing and direction, the wind speed and direction can be deduced; alternatively, from a given wind speed and direction the isobaric pattern can be assessed.

The application of this two-way relationship is always the first step in the drawing up of weather charts. But here allowance has to be made for two other factors: *Surface Friction* and then *Curvature of the Isobars*.

Surface friction
Surface Friction is least over the sea and greatest over the roughest land. Speed is reduced, direction altered, and gustiness for both speed and direction is increased.

A safe general rule to apply for direction is that over the sea surface friction will alter the direction by 5° to 10° (a little under one Compass 'point' or $11\frac{1}{4}°$) and by 10° to 15° over the land. In the Northern hemisphere alteration is counterclockwise or 'backed', and in the Southern, clockwise or 'veered'.

The effect of Surface Friction decreases with altitude so that at 2,000 ft. the wind is parallel to the isobars, and blowing at the speed determined by the Pressure Gradient as calculated from the Isobaric Spacing (as with the Geostrophic wind).

The difference between the wind-speed at 2,000 ft. friction-free, and the actual wind at the standard 33 ft. (10 m.) surface level depends on:
1. The type of air-stream.
2. The time – day or night.
3. The location – land or sea.

Warm, moist (stable) air-streams are most affected by friction and in them the surface wind is reduced to about 65% of the wind at 2,000 ft.

Cold, dry (unstable) air-streams tend to bounce along over irregular surfaces, and especially over warm surfaces. The net result is that the air-stream becomes gusty, both for speed and direction. The speed of the maximum gusts approximates to the Geostrophic Wind.

At night, surface cooling of the Earth means that the coldest layer of air is the one in contact with the surface. After sunset (or even before) this layer can effectively reduce the surface wind to calm. This effect is most marked in the 'Fair to Cloudy' weather associated with cold unstable air-streams.

Over the sea, friction does not have day/night variations; consequently there is no moderation in wind speed in the evening and no freshening after sunrise, as there is over the land.

Curvature of the Isobars, and the Cyclostrophic Force
In practice it has been found that *actual* wind speeds around Lows are less than the values calculated from

the Pressure Gradient as shown by the spacing of the isobars, and that at the very centre of the Low there is a weird unnatural calm. Similarly, winds around anticyclones – Trade winds for example – are *stronger* than the gradient value.

This anomaly is caused by what is known as the *Cyclostrophic* (circle-turning) force. Its effect is shown diagrammatically in Fig. 5, and is positive around anticyclones (Highs), i.e. centrifugal; it is negative around cyclones (Lows), i.e. centripetal.

The Cyclostrophic Force

Table showing how actual wind speeds do indeed vary with changes in geostrophic wind values and in the radius of curvature of isobars. These figures are for Lat. 55°.

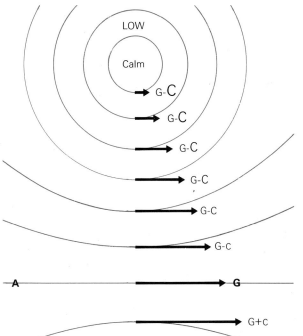

Fig. 5. *Cyclostrophic effect on the wind. The arrow length indicates the speed of the wind. Around a Low (cyclone) the cyclostrophic force C increases with the curvature of the isobars, and its effect is negative, so reducing wind speed. In a High Pressure area (anticyclone) the effect is reversed, C is positive and wind speed increases. Along straight isobars (A) wind has the basic geostrophic value.*

Geostrophic wind speed in knots	Radius of curvature of isobars in nautical miles							
	100	150	200	300	400	500	750	1,000
Actual wind speed (in knots) where curvature is cyclonic								
5	5	5	5	5	5	5	5	5
10	8	9	9	9	9	10	10	10
15	12	13	13	14	14	14	14	15
20	15	16	17	18	18	18	19	19
30	20	22	24	25	26	27	28	29
40	25	28	30	32	33	34	36	37
50	30	33	35	39	40	42	44	45
Actual wind speed (in knots) where curvature is anticyclonic								
5		5	5	5	5	5	5	5
10		12	12	11	11	11	10	10
15		24	19	17	17	16	16	16
20			32	25	23	22	21	21
30				47	39	36	33	32
40					63	53	47	45
50						79	62	58

Cirrus

THESE THIN WISPS of high cloud, blown into streaks by the strong winds of high altitude, look very like 'mare's tails'. Warm, moist Tropical Maritime air, in its general eastwards movement, lifts upwards at the Polar Front over cold, drier Polar air, and even at 20,000 to 30,000 feet, still retains enough water to form these clouds of minute ice-crystals. They are the first and sure indication of a depression, or low, some 500 to 1,000 miles away, to the west.

Note that cirrus cloud like this can be the last dying streaks of large cumulo-nimbus (thunder) clouds, but the sky is then not usually covered.

Navigators are warned that this may be their last chance of position-fixing by astro-navigation for 36 to 48 hours, or even longer, under the overcast conditions of depressions.

Cirro-stratus

HERE THE CIRRUS CLOUDS have grown thicker and lower. The air is now too humid for the water-vapour from the aircraft engines to evaporate, so their condensation trails become persistent. Such persistent trails often precede cirrus cloud, an indication that moist Tropical air is invading the region at great height, in all probability from the warm front of a depression about 1,000 miles to the west.

Stratus and fracto-stratus

THE OVERCAST OF stratus-cloud, its base down to 1,000 ft or even lower, covers a wide area with more or less continuous rain. Below the main cloud base are broken banks of even lower cloud driving along — *scud* to the older mariner — in the Gale Force winds which so often accompany this sky pattern.

Not only is astro-navigation impossible but visibility becomes POOR at 2 miles or less, position-fixing by reference to charted landmarks, i.e. coastal navigation, becomes difficult.

Stratus

AFTER THE PASSAGE of the warm front, and when moist Tropical air covers an area, winds off the sea can create long banks of low stratus along the cliffs and headlands, as shown in this photograph.

In summer, by day, the sun is usually warm enough to 'burn-off' these low banks of cloud but in winter they normally persist and drift inland for several miles.

Fog banks invariably occur along coasts where tidal streams bring cold water to the surface over shoals — water cold enough in spring and early summer to cool the air above it down to dew-point.

Cloud formation

For *Cloud* to form, the air must be cooled to the *Dew-Point* – the point when the water-vapour will condense to water drops or, at high altitude or in high latitudes, into ice-crystals.

If the cooling is rapid enough and if the proportion of water-vapour is high enough, the rate of formation of water drops will be too great for the air to carry them in suspension and they will therefore fall to earth as rain, snow or hail.

The simplest type of cloud formation is when a warm moist air-stream flows over a cold area of land or of sea. The air-stream cools to its Dew-Point, condensation takes place and a cloud is formed – at ground level – as fog (or mist) and fog it will remain unless some other factor, such as wind turbulence, lifts it above surface level to become low cloud.

Apart from this rather special case, cooling to condensation point and the resultant formation of cloud with the related possibility of rain, can occur in three ways.

---

1. Orographic cloud

Orographic is the name given to the type of cloud (and to the rain from it) which forms where hills or mountains are exposed to moist air-streams, forcing the latter to rise, and, rising, to cool to the Dew-point. The cloud form is usually extensive *Stratus* layers low enough to become hill-fog over the higher slopes.

Thus, all coasts exposed to prevailing on-shore winds will frequently have cloudy to overcast weather with regular rainfall throughout the year, or through whatever season these winds do prevail.

2. Convection

Any air-stream passing over heated surface areas will absorb heat from those surfaces, expand, rise and cool. It may perhaps cool to the Dew-Point and so start a cloud. The cooler the air-stream, such as winds coming off the Polar regions, the more marked will this effect be. However, as surfaces sufficiently warm to create uplift are usually relatively small in area, individual *Cumulus*-type clouds are formed, the weather being 'Fair' or 'Cloudy with showers'.

Convection cloud and convection showers can occur anywhere in the world provided there is sufficient heating uplift and the air-stream is sufficiently moist. Such showers are not unknown even in desert areas, such as Arabia or Central Australia, which occasionally experience such rain. Convection cloud formation is most developed and convection rainfall most abundant where the Trade Winds meet near the Equator. The combination of moist air-stream and heat is then of course ideal.

3. Frontal

When a warm, moist air-stream meets a colder, drier air-mass it will, *being lighter in weight*, rise up over the heavier, cold, dry air, just as up a mountain-side. Over the oceans vast areas of *Stratus Cloud* form, swirling round and upwards and moving eastwards, bringing long periods of overcast weather with continuous moderate to heavy rainfall beneath. The boundary between the warm moist air and the cooler, drier, heavier air is described as the *Front*, a term which must be familiar to anyone who listens regularly to the weather forecast on radio or television. This type of cloud is the one which brings the normal, all-year-round rainfall of the temperate Latitudes.

Precipitation

Precipitation is the overall term used to describe what falls out of the clouds in the sky; and it is water, in one form or another.

Precipitation may be widespread or very local, identifiable with one recognisably individual cloud or cluster of clouds.

This precipitation – called a *Shower* or *Showers* – lasts as long as it takes for the clouds to pass over. It may be a few minutes or an hour or two, depending on the geographical area of the cloud and on the speed at which it is moving. A large, very slow-moving cloud may give precipitation for several hours. Nevertheless it is, by definition, a *shower*.

In temperate Latitudes shower activity is a characteristic of Polar air-streams (cold and un-stable) especially when they are subject to uplift over mountains or by crossing sun-heated land by day.

Shower clouds

It is the *Cumulus* (accumulating) family of clouds which generate showers beneath them. They do so when their rate of vertical development is fast enough for the cooling and condensation of water-vapour in the up-currents to be greater than the warming and re-evaporation in the down-currents. A second necessary factor is that the vertical development must be past the freezing level where some drops freeze into hailstones and where other drops become super-cooled (below freezing), and so can coalesce rapidly into the large heavy drops so characteristic of shower rainfall.

When the freezing level is low, clouds of small vertical development will give showers – for example
▶ *Page 26*

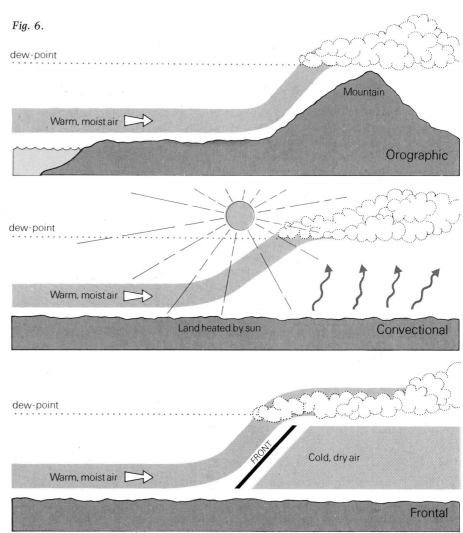

Fig. 6.

dew-point

Mountain

Warm, moist air

Orographic

dew-point

Warm, moist air

Land heated by sun

Convectional

dew-point

FRONT

Cold, dry air

Warm, moist air

Frontal

Low stratus over land

DEEP, SLOW-MOVING, DEPRESSIONS cover large areas with thick sheets of grey stratus cloud, and hill and coastal fog. Navigation for both mountaineers and sailors becomes difficult and hazardous.

This weather is so common between latitudes 45° and 55° (North and South) that lighthouses there have to be sited at the base of headlands rather than on their summits. Vessels frequently have to be dangerously close before landmarks can be identified and bearings taken from them to give an accurate fix of position.

Strato-cumulus

WHEN TROPICAL MARITIME air is not too moist and (as in summer) the sun can warm land surfaces sufficiently to give some convection, a layer of stratus cloud can change to strato-cumulus and gradually break up to give fair or fine periods.

Visibility improves to MODERATE (3–5 miles) or even to GOOD (5–10 miles) and the stars can be seen again.

The barometer will probably reflect the drying of the air-mass by rising slowly.

Cirro-cumulus

THE VERY SMALL quantities of water-vapour at the low temperatures of the higher layers of the troposphere restrict the development of cirro-cumulus and of alto-cumulus to clusters or lines of tiny cloudlets, which have the characteristic fish-scale appearance called 'mackerel sky'.

This is often seen after periods of stormy weather when drier but cold, unstable Polar air replaces Tropical air. The air is drying out, skies are clearing to give fine weather – more sunshine in summer but, alas, more frost or fog in winter.

Above this layer of cirro-cumulus is a man-made cloud, the condensation trail of a jet aircraft.

Cumulo-nimbus

THIS PHOTOGRAPH SHOWS the classic 'anvil' shape of the fully developed cumulo-nimbus cloud. It has violent turbulence within, and water drops can be carried upwards through the freezing-level many times to become large heavy hail-stones.

Sooner or later cold, Polar air swings in from the west behind all depressions, pushing violently under the warm Tropical air ahead of it. The line of the advancing cold, unstable air, the *cold front*, is marked by violent, squally showers often with thunder. These clouds tower up to 30,000 or 40,000 ft. and their bases lower to below 1,000 ft.

the famous April showers in the Northern hemisphere when the freezing level can be as low as 1,000 ft. to 2,000 ft.

In the Tropics, Cumulus clouds can tower to great heights (over 25,000 ft.) before the cloud will break into a tropical downpour, most certainly accompanied by thunder. With vertical development of over 12,000 ft. the turbulence within any *towering Cumulus* will generate electric charges and thunder can be expected.

Cumulo-Nimbus is the correct technical name for a cloud from which there is precipitation, *Nimbus* being the Latin name for rain. Once precipitation has started the rapid condensation of water-vapour to water drops releases latent heat further to warm the air and keep the uplift going until either all the water-vapour becomes used up or the cloud rises to some air above it of a higher temperature (another air mass) through which it is not warm enough to rise. At this point the cloud-top levels off to give the anvil-head of the fully developed Cumulo-Nimbus cloud.

Wind is the most important factor to consider when attempting to assess future weather conditions. This is because it is the wind which actually brings us the weather we experience.

For example, winter in the Northern hemisphere often brings a period, lasting several days, during which north winds blow quite severely. It will then generally be cold, but conditions will still vary from area to area. Mountainous country to the north will probably be very cold with snow showers as the wind, lifting over the mountains, cools to form cloud and swirling blankets of hill-fog. Then, if the process is vigorous enough and continuous enough, the clouds will give blizzards or driving rain – but every

snow-flake or rain drop which falls is one less for that air to carry southwards. Every southward mile is a mile towards warmer land. Every mountain range or gentle hillside crossed probably adds another subtle degree or two to the temperature.

Thus, as it progresses southwards that same north wind, its bitter sting removed, becomes quite tolerable to the human frame. However should the winter winds settle into the east, as so often happens, the south of a country can be as cold as the north and possibly colder!

The actual surface temperature in any locality largely depends on the track by which the air arrived. For example, air reaching North East Scotland from the east across the North Sea will, in its lower layers, acquire a temperature very closely related to that of the sea over which it travels: quite possibly a few degrees *above* freezing point. In marked contrast, an east wind can reach South East England at temperatures well *below* freezing point after travelling several hundred miles over the frozen snow-covered plains of Poland, Germany and the Low Countries. Such east winds have been known to persist almost without a break from November through to March. Most unpleasant!

Not surprisingly, it is the south westerly winds which bring most warm air to the eastern United States. Blowing intermittently – maybe a few hours, occasionally for a few days – they bring warm air from the sub-tropics. However, although this warmth is clearly recorded by thermometers, our bodies react to the high humidity of these south-western winds and tell us that the weather does in fact not *feel* as warm as the thermometer suggests. Moreover, this high humidity creates sufficient cloud cover to shut out the warmth from the sun, and day temperatures tend to remain low. That minimum night temperatures remain high – to keep the daily average up – is no consolation to our human senses which tell us that the weather is *'damp and cold'* in spite of the evidence of the thermometer.

These examples indicate in a very general way how the *Weather* over any given geographical area, north or south of the equator, depends on where the air has come from. The next factor which affects it is the wind speed.

Wind Speed

A wind of Beaufort Force 4 is described as a 'Moderate Breeze', not light enough to be negligible nor strong enough to be a nuisance. It will just keep the smaller branches of trees in motion; at sea the waves will be small but some crests will break into 'white horses'. The speed of the Force 4 is between 13–18 m.p.h. or 11–16 knots (nautical miles/hour).

Now let us do a few simple sums, using knots as units rather than statute miles per hour, because 60 nautical miles is 1° of Latitude.

Taking a mean speed of 15 knots for a Force 4 wind:
In 6 hours   the air moves   90 nautical miles
In 24 hours the air moves 360 nautical miles =
(1 day)   6° of Latitude
In 48 hours the air moves 720 nautical miles =
(2 days)   12° of Latitude
In 72 hours the air moves 1,080 nautical miles =
(3 days)   18° of Latitude

So a stream of air moving southwards off the drifting ice near the Arctic Circle, Latitude 66½° North, at about 15 knots, will reach Winnipeg, Canada, Lat. 50°N, in about two and one-half days and New York City, Lat. 41°N in a little over three days.

This air stream will probably warm up somewhat on its way south, especially in summer when long hours of sunshine are possible – over 20 hours – and night cooling is reduced to a few hours. In winter, however, darkness covers these northern latitudes for most of the 24 hours and even Chicago has only about 11 hours of daylight.

East Winds

East Force 4, Speed 15 knots, 360 n. miles per day or approximately 12° Longitude.
Berlin (Long. 13°E) to Greenwich, England,
(0°E/W) – about 1 day
Lvov, Ukraine (24°E) to Greenwich –
about 2 days.
A steady moderate easterly breeze can bring to north eastern England the bitter but clear, dry air from the frozen steppes of the Ukraine in just about 2 days. The same wind in summer will bring dry heat but a dust-laden atmosphere of poor visibility.

West Winds

West winds bring much the same conditions of cloud and of temperatures whether they are Force 4 or Force 7. The waters of the North Atlantic ocean are warmed by the Gulf Stream-cum-North Atlantic Drift and the speed at which the air moves towards north east Europe does not matter very much. What does matter is where that air came from before it arrived over the mid-North Atlantic. The answer will be found in Chapter 3.

Towering cumulus

THE WEATHER IS still showery but the shower clouds are less violent and the tops are levelling off at a lower height. The wind is less strong and less gusty.

Visibility is always good in this clear Polar air, except, of course, in the showers. Navigation becomes easy once again as the depression moves away eastwards.

The wind will be moderate to fresh for some time, and gusty for both speed and direction: over land the wind and the showers both die away in the late afternoon, to return again the following forenoon.

Fair weather cumulus

THESE ARE PROBABLY the most beautiful of all the clouds brought by light to moderate, generally westerly, winds. The Polar air is clear and cool and the humidity relatively low at 50%–60%, giving a $\frac{3}{8}$ cloud cover. All the clouds have well-defined cloud base, usually between 1,500 and 2,500 ft., as the air lifts by convection from a land surface warmed by the sun climbing up during the forenoon.

If these clouds appear early in the morning they can be expected to develop into big shower clouds. If they do not form until the late forenoon there will be little change until they die away during the afternoon to be followed by a clear star-lit night. Out of the wind, even in winter, the full warmth of the sun can be felt and appreciated.

Cumulus over the sea

THIS PHOTOGRAPH TAKEN in Jamaica quite early in the day shows that the sea is warm enough to create convection and convective cloud over the water. By contrast the land is still too cool but as it heats up the clouds, already well-developed over the sea, will be drawn in over the land by the sea-breeze and then erupt into the violent afternoon thunderstorms which characterise all islands in the Tropics.

In temperate waters, such as the English Channel, sea temperatures remain much higher than land temperatures throughout the winter and into early summer, and cumulus clouds develop in Polar air streams both by day and by night, in contrast with the land where they only develop by day.

'Red sky at night'

THE MOST GLORIOUS SUNSETS, and sunrises, are seen when the sun is setting, or rising, through extensive banks of stratus or strato-cumulus cloud before they are too thick and too low to shut out the light of the sun altogether.

These cloud banks are almost always associated with an approaching warm front, in turn associated with generally 'foul' weather, not immediately but in some 12 to 20 hours' time. 'Red in the Morning' is the seaman's warning of a foul night ahead of him. 'Red Sky at Night' – delights him with the prospect of a night's rest before the storm he can expect during the following *daylight* hours.

Chapter 3
How air masses determine the weather

WHEN IN MOTION, air moves in what we know as *Air Masses* — a term which indicates the sheer volume of air involved. The two main air masses are *Polar* and *Tropical*, defined as such by their source area. As these masses move, they track either over the sea to become *Maritime* or over land to become *Continental*.

Polar air

The great frozen expanses of the Arctic and the Antarctic are the great 'air-conditioners' of the world.

However warm, wet or dirty with industrial pollution air may be when it reaches Polar regions, once there the steady cooling condenses out almost all the water-vapour as rain or snow, thus removing all the dirt and impurities, and depositing them on the frozen surface below. Polar air is clean as well as cold.

The Polar areas generate an endless supply of Polar air — cold, clean almost completely dry, and, therefore *heavy*. It can spread in great, cold refreshing waves across the Temperate Zones of the world, at times surging down into the Tropics. As it moves southward off the Arctic towards Europe or N. America, or northwards off Antarctica towards Australia, New Zealand or South Africa, it travels over progressively warmer regions. The lower layers in contact with the surface of the sea or land warm up first, expand, become lighter, and rise upwards through the colder layers above. As this air goes up, other air must come down to take its place. The air stream becomes turbulent, and in a condition known as unstable. As it begins to move, according to its track it becomes either *Polar Maritime* or *Polar Continental* air.

Polar Maritime air

This is cold to start with, and with little water-vapour content. However, as it moves towards more temperate latitudes it passes over seas progressively warmer and picks up water-vapour at a progressively faster rate. Thus the lower layers become not only warmer but moister as well.

Now, wet air, strange as it may seem, is lighter than dry air. These warm, moist, lower layers are thus too light in comparison with the cold, drier layers above. Polar Maritime air is therefore, even more unstable than Polar Continental air, and more local upward swirls are generated within it, often producing towering cumulus clouds. When this type of cloud has rapid and very excessive vertical development it generates within itself thunder, lightning and violent turbulence.

Polar Continental air

This is cold, relatively dry, and with a tendency to become unstable as it moves away from Polar regions over progressively warmer land surfaces. However, there are very marked differences between its *Winter* and *Summer* characteristics.

Winter: The interiors of the two large land masses of the world — North America and Euro-Asia — become very cold and very dry in Winter. The days are short; the nights long. The heat gain by day is progressively less than the heat loss by night and the process is *not* reversed on mid-Winter's day, 21st to 22nd December. The coldest four weeks in the Northern hemisphere are mid-January to mid-February, for it takes some five to seven weeks for the days to lengthen appreciably and for the sun to climb high enough in the sky to give more warmth by day than the heat loss by night.

In the Southern hemisphere there is, apart from the ice-covered land mass of Antarctica, no land south of Lat. 50°S, other than the tip of South America, and a few remote islands. Thus no Polar air reaches lands such as Chile, Argentina, Australia and New Zealand, without a long sea crossing which turns such air into Polar Maritime.

Polar Continental air, therefore, is only experienced in the temperate latitudes of the Northern hemisphere – Canada, U.S.A., Europe and Asia.

Any air moving over these cold expanses, between November and even as late as May, will remain cold, dry and stable. For much of the time the lowest layers in contact with the surface will remain very cold and heavy, trapping within themselves industrial dirt and pollution. This smog will concentrate in valleys and low-lying areas unless wind strength is great enough to dissipate it.

Summer: Over the vast land masses of the Northern hemisphere lying between the 50° Parallel of Latitude and the Arctic Circle (Canada and U.S.S.R.), the long mid-Summer days lift daytime temperatures well above 25°C (80°F). These high temperatures at low levels, combined with low temperatures at altitude make the air unstable. The warm air, rising as vigorous 'thermals', produces cumulus clouds. Passing over the rivers, lakes and moist surface soil of Canada and U.S.S.R.,

it absorbs more moisture, so creating *towering cumulus* clouds, many of which develop into violent thunderstorms.

Polar air, tracking southwards over this terrain becomes warm, relatively moist, unstable and showery. It can extend this type of weather southwards almost to the Tropic of Cancer, bringing cool, refreshing rain to lands which otherwise would be hot, dry deserts.

This effect, however, is limited to the late Spring and early Summer. As summer advances the transfer of heat upwards throughout the entire land mass makes the Polar air-stream less dense, less heavy. The pressure becomes low.

As a result, other air-streams are drawn in over the Continental land masses – a process particularly well-developed over South-East Asia and known there as the South-West Summer Monsoon.

Tropical air

If the Polar Regions can be regarded as the two great *'Air-conditioning'* areas for the earth's atmosphere, the Tropics are the great 'central-heating' areas. Within the Tropics, from Cancer across the Equator to Capricorn, day and night are nearly equal throughout the year. The heating by day and the cooling by night are similarly balanced and it is always hot. The daily average temperatures remain steadily high. The path of the rising sun is a steep climb upwards in the east. Within an hour it is high enough and strong enough to cancel the heat-loss of the night. Although the downward path of the sun to its setting in the west is as steep as its morning rise, not until several hours after sunset do surface temperatures fall appreciably.

Surprisingly, really intolerable heat is found,

▶ *Page 34*

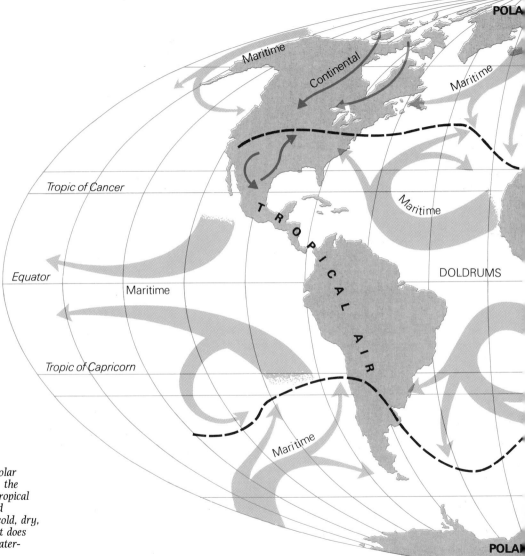

Fig. 7. In both Northern and Southern hemispheres Polar and Tropical air meet somewhere in the Temperate latitudes. At that point there is a boundary line — a frontier between them — which is called the Polar Front. At this Front, the warm, moist, light Tropical air rides upwards and pole-wards over the cold, dry, heavy Polar air. As it does so, it cools and its water-

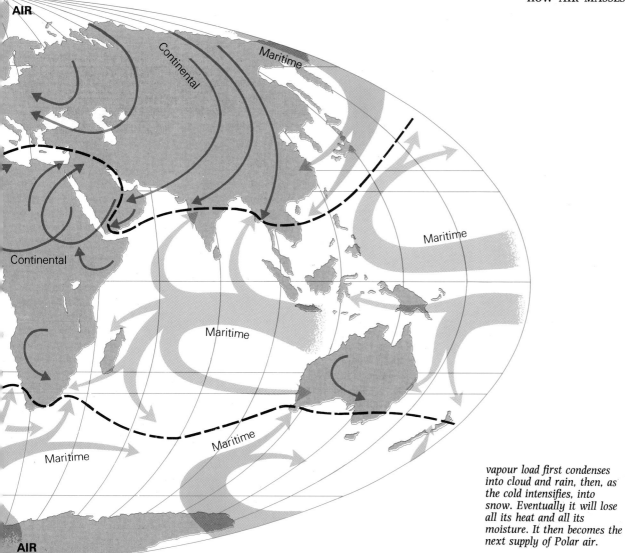

AIR

Continental

Maritime

Continental

Maritime

Maritime

Maritime

Maritime

Maritime

AIR

vapour load first condenses into cloud and rain, then, as the cold intensifies, into snow. Eventually it will lose all its heat and all its moisture. It then becomes the next supply of Polar air.

Similarly, the surges of cold Polar air across the Temperate zones gradually warm up, absorb moisture and enter the swirls of the tropics to become the next supply of Tropical air.

not on or near the Equator but between Latitudes 15° and 30° North and South of Equator and well away from the sea – in the hot desert lands of the world, which all lie on the Western sides of land masses.

In North America: parts of California and Mexico
In South America: the Atacama Desert
In South Africa: the Kalahari Desert
In North Africa: the Sahara
In India: the Desert of Sind
In Australia: the Central Desert

Tropical Continental air

By its very nature, any air mass remaining for long over these tropical land masses will become very hot, even up to great heights, and it will be light. At the surface more air will be drawn in to replace it as it rises, and this cool replacement air will remain at low levels until it also heats up and rises. Eventually the air heated in this way will move away from the Tropics at great height, to spread its heat energy towards the Poles.

As a general rule, Tropical Continental air tends to rise very high and not to affect the weather of adjacent areas. However, with marked seasonal variations in their frequency, hot, dry, dust-laden winds such as the *Sirocco*, *Khmain* and *Haboob* do occur locally to give some substance to the term *Tropical Continental Air* and the weather associated with it.

Tropical Maritime air

Over the oceans in the Tropics, the sun's rays penetrate the sea to a great depth, spreading the heat energy throughout a vast volume of water. There is no great heat-loss from the surface by night.

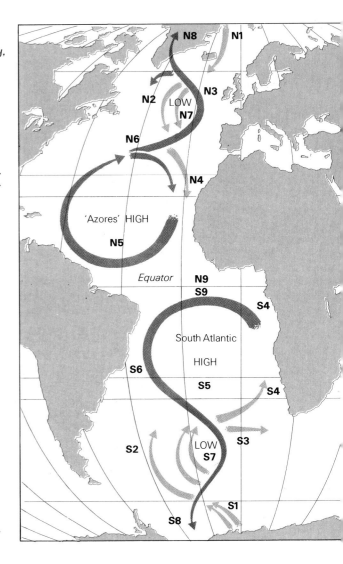

Fig. 8 Air flow over the North Atlantic.
N1: Polar air, very cold, dry, heavy and stable. Pressure high, weather fine, clear, dry. Wind N–NE.
N2: Polar Maritime air. Warmer, moister, unstable. Cloudy, snow showers, otherwise clear.
N3: Tropical Maritime. Warm, moist, light, rising over Polar air. Pressure low. Cloudy/overcast. Rain/snow. Strong wind.
N4: Polar air, now warm, mixing well with cooling Tropical air. N.E. Trade Winds, fine, warm.
N5: Areas of almost permanent high pressure called the 'Sub-tropical Oceanic Highs'. Generating area for Tropical Maritime air. Fine, warm. Winds decrease to light, variable at centre.
N6: Area of N. Atlantic 'Westerly' winds.
N7: Breeding area of Temperate zone cyclones or Lows (counterclockwise) swirls, between lat. 40–60°N.
N8: Tropical Maritime air now cooling and sinking. Last remaining water-vapour falls as snow. It is becoming Polar air. Unsettled, cloudy, fogs, blizzards.
N9: The doldrums. Very hot, very wet, very light air. See S9.

Tropical seas become a great store of warm water which, although tending to be salty (and so heavier) due to surface evaporation, remains light enough to float over colder, fresher water when ocean currents move it to temperate latitudes.

The air over the oceans will never become warmer than the water (about 80°F), but instead becomes extremely humid.

For example, 10°C rise in temperature from 20°C to 30°C almost doubles the maximum possible water-vapour content (from 15 to 27 grams per kilogram); a further 10°C rise almost doubles it again.

Thus Tropical Maritime air not only carries considerable heat energy, it also bears a vast load of water-vapour. The moment this moist tropical air moves from its source area it will inevitably become cooler. This leads to condensation of the water-vapour into the water-drops of the cloud. If the process is rapid enough, the water-drops will form too rapidly and too large in size to remain airborne. Rain falls. Tropical Maritime air, is therefore, the great rain-bearer of the world.

Tropical Maritime air is a carrier of heat in two ways: firstly by its own heat capacity, and secondly by carrying the latent heat contained in the water-vapour content of the warm moist air.

Just as Polar air surges equator-wards in great tongues of cold, clean, invigorating air, so Tropical air, especially Tropical Maritime air, swirls outwards from the Tropics to spread warmth and to drop its load of water-vapour as rain over the Temperate zones and as snow over Arctic regions.

The general tracks of these Polar surges and Tropical swirls are shown on the map on page 34.

How an Air Mass moves

Imagine a 'parcel' of air 20 to 30 miles (32–48 km.) across and about 5 miles (8 km.) high, consisting of cold, dry, heavy Polar air. It is situated to the north of Greenland and then starts moving as part of a surge towards the equator. It picks up speed, covering about 360 nautical miles a day. Travelling over the warmer waters of the North Atlantic, it warms a little and picks up moisture; it has thus now become Polar Maritime air. The 'parcel' swings over the British Isles, France and Spain, giving those areas a few hours of crisp, cold weather with bright sunny periods broken by short squally showers.

By the time it reaches North Africa it has lost its cold bite. It is warm enough to be absorbing moisture instead of precipitating it. Its path now swings towards the equator and it is part of the famous north-east trade winds, becoming Tropical and also mild to warm. A few days later the 'parcel' is moving westwards or is even becalmed in the Doldrums. It is now Tropical Maritime in character. The total distance it has covered is approximately 7,000 miles.

Sooner or later it will start to move westwards, then north and then north-eastwards. It will become part of the North Atlantic 'westerlies'. Somewhere near the British Isles it will meet and ride up over Polar air on its way down from the Arctic. Its water-vapour content then condenses into vast sheets of thickening, lowering clouds, before spreading rain northwards and eastwards over the British Isles. Moving on towards Norway, its rain will turn to sleet and then to snow. Eventually the last drops of moisture will be chilled out of it as blizzards, somewhere beyond Spitzbergen. It is back where it started.

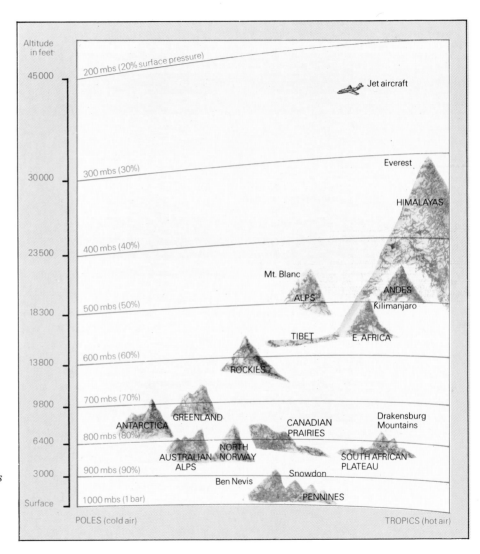

Fig. 9. How pressure lessens with altitude, as the atmosphere becomes progressively thinner and lighter, and the weight of air above is reduced.

Chapter 4 Up into the atmosphere

THE ATMOSPHERE IS a thin layer of *air* which envelops the earth and gives it those special qualities which set it apart from the other planets in the solar system.

What is air?

Air is a mixture of oxygen and nitrogen in the proportions of about 20% oxygen and 80% nitrogen. These proportions are not exact, but are easy to remember and accurate enough for our purpose here. The important fact about these percentages is that they are always the same wherever a sample of air may be taken for analysis. The reason is basically that oxygen and nitrogen are almost the same weight as each other. Oxygen atoms weigh 16 units each and nitrogen 14 each and both exist as 2-atom molecules, O_2 at 32 units and N_2 at 28 units respectively. So long as the earth spins and the sun shines and the atmosphere is in constant motion oxygen and nitrogen remain well mixed in these 20/80 proportions.

In addition to oxygen and nitrogen, there is some carbon dioxide present. All animals breathe in air, absorb some of the oxygen and breathe out carbon dioxide in its place. Yet all this carbon dioxide, natural and man-made, only totals 0.03% of the air or 3 parts in 10,000. As fast as animals breathe it out, plants absorb it, using the carbon for their growth and putting oxygen back into the air.

In addition to carbon dioxide, there are minute traces of other elements grouped together as the 'Rare Gases'. Of these the best known are helium, the lightest of all gases after hydrogen, and neon, the 'new' gas which has achieved fame because it glows brightly when encouraged to carry a small electric current.

As with oxygen and nitrogen, the quantities of these rare gases remain steady the whole world over, from Equator to Poles and from sea level to the limits of detection. Helium is the exception. If any is made and permitted to escape it will rise, and go on rising, into space.

Amongst the constant company of air molecules is an invading army of a different sort of molecule, a restless, energetic, eccentric molecule. The chemical is known to be *Hydrogen Oxide* (H_2O), otherwise water. Its molecular weight is 18 units; 2 of hydrogen (whose weight is 1 unit), plus 16 of Oxygen. So in

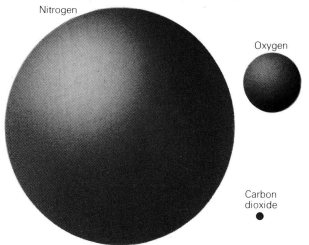

Nitrogen

Oxygen

Carbon dioxide

Fig. 10.

gas form, water is lighter than either oxygen at 32, or nitrogen at 28 units, and lighter than carbon dioxide (CO_2 of 14 and 32) at 46 units.

Now if water-vapour is lighter than air why does it not float up to the top of the atmosphere and disappear into outer space? The answer is that it would if it could, but it can't. It has to be heated to make it a gas and if it cools it changes back into liquid water again. Water-vapour in the air rises but, rising, cools: cooling, it condenses back to a weighty droplet and falls back to the surface, energy-less.

The water molecule is the heat carrier of the atmosphere. Absorbing heat received from the sun it can store that heat, change state from liquid to gas and leap out of its own kind into the atmosphere above. Initially the air aids this process. The warmer the air the more water-vapour molecules it can hold. The faster the air is moving, the faster this process of evaporation goes on. Sooner or later however the air turns hostile and expels the water-vapour, robbing each water-vapour molecule of its hidden heat and dropping it back to earth as rain.

Where the air makes this sudden and very large gain in heat may be a thousand miles or more from where it picked up the water from the surface of the earth or the sea.

Wisps of cloud very high in the sky (*Cirrus* or 'Mare's Tails') are witness of a double robbery – first when the air took the heat from the water-vapour and turned it to water drops, secondly when it took a further measure as the drops froze into ice-crystals.

The upper limits of cloud and therefore, of 'weather' is somewhere between 5 and 10 miles (8,000–16,000 m.): about 5 miles (8,000 m.) in the dense, cold air over the Poles, rising to 10 miles (16,000 m.) in the warmer moister air of the Tropics.

The troposphere

This is the name given to this lower layer of the atmosphere. *Tropos* is Greek for 'turning' and by inference it well describes the rolling, swirling energy of the total enormous weight of the atmosphere, in constant and often violent motion over the face of the Earth.

Within the troposphere the atmosphere thins off rapidly, the temperature falls-off rapidly and so does the amount of water-vapour. The fall-off in temperature eventually ceases, and this occurs, not surprisingly, at about the same altitude at which the water-vapour content has become, for all practical purposes, nil.

Tropopause

This is the name given to the dividing line between the troposphere and the layer above.

The stratosphere

The layer above the tropopause where temperature remains almost constant at about $-60°C$ is known as the *Stratosphere*. It extends to about 40 miles up (65,000 m.). This layer of even temperature prevents thermal uplift currents of air rising either into it or through it unless such uplift is exceptional, as in the case of tropical thunderstorms and the mushroom clouds of nuclear explosions. The total energy of both these is remarkably similar, the big difference being that the nuclear cloud has been given a much greater initial boost than nature can give in a thunderstorm.

The stratosphere is the operational medium for jet aircraft. The temperature is uniform and low which gives a steady, high efficiency to the engines, visibility is good and there are none of the usual weather hazards of the troposphere. In general the winds encountered are moderate in speed and steady for both speed and direction – in general, but not always.

Locally, *Jet streams* can however occur, usually just within the stratosphere. Oval in cross-section, a few miles across (4,000–5,000 m.) and less than a mile (1,500 m.) deep these jet streams snake eastwards right round the world. Fortunately there is a relationship between their tracks, the wind circulations and the position of the Polar front between Polar and Tropical air. As a result their formation and position can be anticipated then checked, so that aircraft can avoid or use them at will. There is nothing to see but entering one submits aircraft to very considerable clear-air turbulence.

Mountains and the weather

The great mountain ranges of the world block or divert the air streams and swirls of the constantly moving atmosphere. They act as effectively as a dam across a river, and as such are an important factor affecting the weather over a wide area. See Fig. 9.

Consider Calcutta and Mount Everest. In March and April, at the end of the winter (north-east) monsoon, the greater part of the Indian sub-continent is very dry and, at sea level, very hot. A balloon ascending from the region of Calcutta would experience a fall in temperature at a very high rate of about 37°F (3°C) per 1,000 ft. (305 m.). Thus, from a surface temperature of about 105°F (40°C), the freezing level (32°F/0°C) would be reached at 13,000 ft. (4,000 m.) and by the time the balloon was on a level with the highest Himalayan peaks (28,000–29,000 ft./8,000 m.) the temperature would be −58°F (−50°C). Paradoxically the balloon-

ists, as with mountaineers at this altitude, would be exposed to the risk of sunburn and snow blindness. Neither could survive for long without either oxygen or some means of warming the air for breathing.

Similarly, if the surface temperature in North Italy were 77°F (25°C), Mont Blanc summit would be about −31°F (−35°C).

On this basis unless the general winter temperature of Scotland falls well below 50°F (10°C) to nearer 41°F (5°C), Ben Nevis and the slopes of the Cairngorms will be on the wrong side of the freezing level for the winter sports enthusiasts. See Fig. 11(a).

This exercise of comparing the temperature at high Altitude with sea level temperature, as for Mount Everest, Mont Blanc and Ben Nevis, can be applied to the other high altitude areas of the world. In Dry Air conditions extremely low temperatures can be expected – and do occur – at night. By day the cloud-free clear skies give maximum heating from the sun and the daily average becomes tolerable. These are the conditions experienced in the inland highland areas of the continents. Examples are the high prairie lands of Canada, the Intermontaine plateaux of the Western States of U.S.A. and Tibet.

However, as we have seen, hot tropical air masses tend to produce great towering thunderstorm clouds, carrying enormous quantities of water vapour, and therefore great amounts of heat to as high as 60,000 ft. (18,000 m.). This could produce a situation where the surface temperature is 40°C (102°F) and the cloud base is about 3,500 ft. (10,000 m.). Up to this height the air is *not* saturated and the temperature will be, at 3,500 ft. (10,000 m.) 80°F (27°C) – the Dry Air rate. At cloud level, however, the air becomes saturated, and the cloud boils upwards, as do all thunderstorms, with a

spectacular release of heat energy. Eventually this cloud will tower above Mount Everest, its temperature falling off at a much slower Saturated rate because of the amount of heat carried upwards. Freezing level will not be reached until over 24,000 ft. (7,300 m.). In Dry air it would have been reached at less than 13,000 ft. (4,000 m.). See Fig. 11 (b).

At 30,000 ft. (9,000 m.) its temperature would be 14 °F (−10 °C) 10 °C below freezing. Not until about 60,000 ft. (18,000 m.) would the temperature fall to −58 °F (−50 °C). Temperatures in the 23 °F (−5 °C) to 50 °F (10 °C) range at altitudes of 25,000–30,000 ft. (7,500–9,000 m.) in these conditions, are tolerable to mountaineers – tolerable, that is, compared with the −58 °F (−50 °C) recorded when the air is dry during March, April and into May. Such moist, warm conditions occur when the hot, wet air stream of the south-west monsoon piles up against the Himalayas in June, July and August. Unfortunately, the monsoon not only raises the temperature, it brings heavy snowfalls, preceded by both avalanches and rock-falls caused by the increase in warmth.

High altitude climbing in the Himalayas is accordingly restricted to a few weeks of May into early June. The climbers ascend just ahead of the monsoon, taking a calculated risk that the clouds and snowfall will not overtake them. A factor usually, but not always, operating to their advantage is that just ahead of the monsoon the atmosphere becomes relatively quiet and the risk of impossibly high winds is minimal.

Some of the Himalayan peaks such as Nanga Parbat, Nanda Devi and Kanchenjunga experience lower temperatures, stronger winds, more severe avalanches and rock-falls because they are in the outer ridges of this complex of parallel ranges and so more exposed to the earlier surges of the advancing South-west Monsoon.

There is a second short period when attempts on the high peaks of the Himalayas are also possible. This is at the end of the South-west monsoon and before the cold, dry, north-east monsoon of the winter months sets in. The 1975 British Expedition made the first ascent of the South face of Everest at this time of year.

Those interested in mountaineering in other parts of the world should bear in mind that maritime air will bring higher temperatures but greater risk of snowfall and avalanches; Continental air has the opposite effect.

Humidity

The word *humidity* simply means dampness. The higher the humidity, the less the air has to be cooled in order for condensation to take place, i.e. for cloud, fog or dew to form. Since cloudiness and rainfall are the major factors in any weather pattern, and these occur when humidity is 100%, knowledge of humidity values, and of dew-point values too, are essential to accurate forecasting. And these values must not only be for the earth's surface, but also upwards into the atmosphere to the limits of weather — the tropopause.

Humidity expressed as a percentage is called *Relative Humidity*. This represents the amount of water-vapour *actually* present in the air at a given temperature, compared with the amount of water-vapour which would have to be present to saturate that same air at that same temperature.

Saturation = 100% relative humidity.

Dew-point

The Dew-point is that temperature at which the relative humidity *becomes* 100% (i.e. the air is saturated) and condensation accordingly takes place, invisible water-vapour changing into visible water-drops. It is the temperature at which an invisible air-stream will breed clouds.

Dry- and Wet-bulb thermometers. In the Stevenson Screens (see Glossary, page 15) on all weather stations will be found a pair of thermometers — a dry-bulb and a wet-bulb. Identical in every respect except that the bulb of one is kept wet by means of a wick from a reservoir of distilled (pure) water, they together make a *Psychrometer* or 'coldness meter' (*Psychro* is Greek for 'cold'). The evaporation of the water round the wet-bulb cools it. The drier the surrounding air the greater that evaporation will be, and therefore the greater the lowering or *depression* of the wet-bulb temperature. By comparing the two readings it is therefore possible to calculate the relative humidity. There is no simple direct relationship between wet- and dry-bulb temperatures and dew-point or humidity, but *Hygrometric Tables* are available to meteorologists from which they can ascertain the dew-point or relative humidity for given wet- and dry-bulb readings. *Hygro* is the Greek word for dampness. See Fig. 11(c).

The more important value is that for the dew-point, since from it one can predict the height at which cloud will form. One use of the weather balloon (radio-sonde) is to relay back to the ground reports on humidity values at various heights up through the atmosphere, and from these reports it is possible to assess dew-point temperatures on which so much else depends.

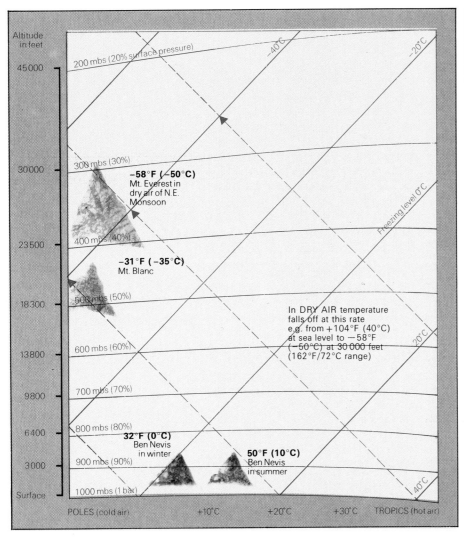

Altitude in feet

45000 — 200 mbs (20% surface pressure) −40°C −20°C

30000 — 300 mbs (30%)

−58°F (−50°C)
Mt. Everest in
dry air of N.E.
Monsoon

23500 — 400 mbs (40%) Freezing level 0°C

−31°F (−35°C)
Mt. Blanc

18300 — 500 mbs (50%)

In DRY AIR temperature
falls off at this rate
e.g. from +104°F (40°C)
at sea level to −58°F
(−50°C) at 30 000 feet
(162°F/72°C range)

13800 — 600 mbs (60%) 20°C

9800 — 700 mbs (70%)

6400 — 800 mbs (80%)

32°F (0°C)
Ben Nevis
in winter

50°F (10°C)
Ben Nevis
in summer

3000 — 900 mbs (90%)

Surface — 1000 mbs (1 bar) 40°C

POLES (cold air) +10°C +20°C +30°C TROPICS (hot air)

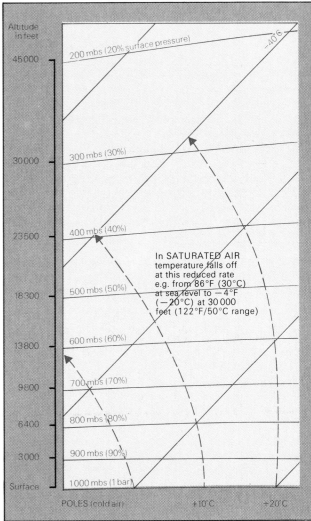

Altitude in feet

45000 — 200 mbs (20% surface pressure) −40°C

30000 — 300 mbs (30%)

23500 — 400 mbs (40%)

In SATURATED AIR
temperature falls off
at this reduced rate
e.g. from 86°F (30°C)
at sea level to −4°F
(−20°C) at 30 000
feet (122°F/50°C range)

18300 — 500 mbs (50%)

13800 — 600 mbs (60%)

9800 — 700 mbs (70%)

6400 — 800 mbs (80%)

3000 — 900 mbs (90%)

Surface — 1000 mbs (1 bar)

POLES (cold air) +10°C +20°C

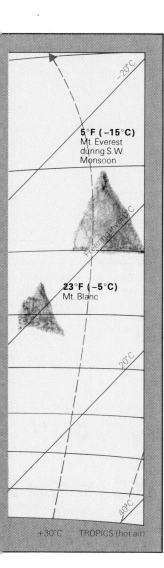

−20°C

5°F (−15°C)
Mt. Everest
during S.W.
Monsoon

Freezing level 0°C

23°F (−5°C)
Mt. Blanc

20°C

40°C

+30°C   TROPICS (hot air)

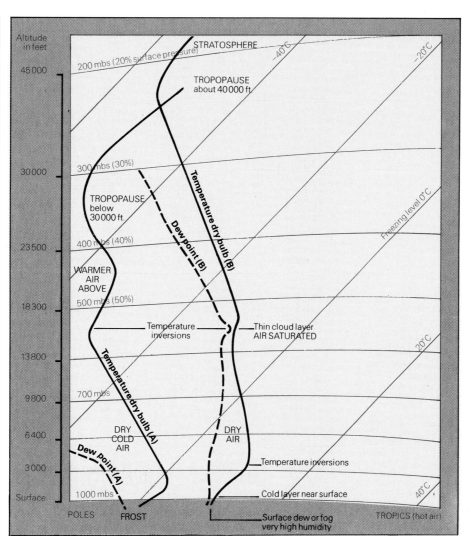

Altitude
in feet

STRATOSPHERE

200 mbs (20% surface pressure)

−40°C

−20°C

45000

TROPOPAUSE
about 40000 ft.

30000

300 mbs (30%)

Temperature dry bulb (B)

Dew point (B)

Freezing level 0°C

TROPOPAUSE
below
30000 ft.

400 mbs (40%)

23500

WARMER
AIR
ABOVE

500 mbs (50%)

18300

Temperature
inversions

Thin cloud layer
AIR SATURATED

13800

Temperature dry bulb (A)

20°C

700 mbs

9800

DRY
COLD
AIR

DRY
AIR

6400

3000

Dew point (A)

Temperature inversions

Cold layer near surface

1000 mbs

Surface

40°C

POLES   FROST

Surface dew or fog
very high humidity

TROPICS (hot air)

Fig. 11. Here are three
'aerological' diagrams, so
named because they give a
cross-section view upwards
through the atmosphere.
11a. This shows how
rapidly the temperature of
dry air falls off, from
104°F (40°C) (at sea
level) to −58°F (−50°C) at
30,000 ft. Upper air is very
cold and heavy, the air mass
top-heavy and unstable —
ideal for violent thermal
up-currents.
11b. This shows how the
temperature of wet
(saturated) air falls off
much less rapidly. Latent
heat is carried upwards in
the water-vapour, and
released as condensation
takes place. It is this process
by which warmth is spread
upwards in Tropical
Maritime air masses.
11c. This diagram shows in
graph form the readings
from two separate Radio-
Sonde ascents: (A) at
midnight, 5/6th December,
1961, and (B) on 11/12th
December, 1961.
Instruments carried upwards
by small balloons measure
values for pressure, dry-bulb
temperature and dew-point,
which are then plotted. From
this data it is possible to
assess what air masses are
present, e.g. (A): extremely
dry with very cold arctic air.
(B): warm, moist Tropical
Maritime air.

Frost and Ice

TEMPERATURES FALL AWAY the higher the altitude, this rate of fall depending on factors explained earlier in this chapter. At the altitude at which surface temperature drops to 32°F (0°C) – and 1,000–2,000 ft. (300–600 m.) above that level – aircraft can be in danger from the formation of ice. Apart from its weight, this ice can ruin the efficiency of the wing shape, or choke off the fuel supply if it affects the carburettor. In the layers just above the freezing level, super-cooled water drops can exist which will freeze almost instantaneously on any intruding air-craft. This risk is greatest in the high humidity of Tropical air masses.

Down at the surface, however, the problem is to assess what will happen if and when the air temperature falls to the dew-point. If in fact there is little or no wind and the temperature is above freezing point, then dew will form at ground level.

If, however, the temperature at the surface falls to 32°F (0°C) or below, ground frost will result. If the temperature *above* ground level i.e. among fruit trees, buildings etc., also falls below freezing point, then air frost will form. In the case of air frost local variations in the radiating properties of differing surfaces become predominant factors.

Fog. If a whole layer of air is cooled to the dew-point, and a gentle Force 1 wind is present, the latter will gently stir all the water-droplets in the air into an opaque 'porridge', which will effectively stop many human activities. A Force 2 wind will tend to lift the fog clear of low-lying areas and the result will probably be patchy hill fog.

When fog and below-zero temperatures combine, the result is freezing fog – the motorist's nightmare.

Chapter 5 Air flow around the world

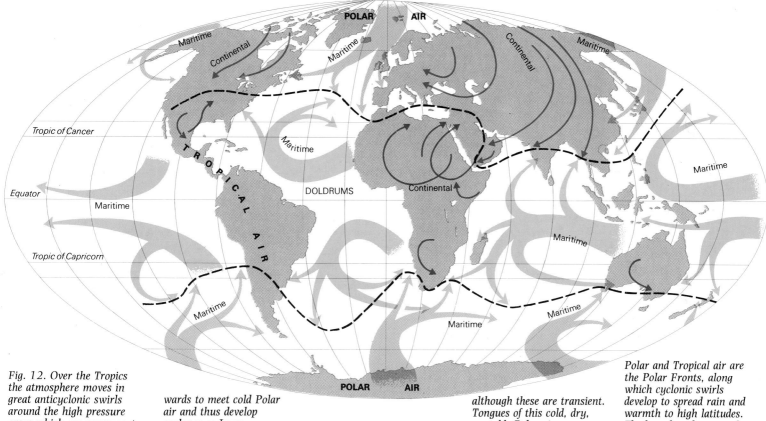

Fig. 12. Over the Tropics the atmosphere moves in great anticyclonic swirls around the high pressure areas which are permanent features over the world's oceans. Tongues of this warm, moist Tropical Maritime air push pole-wards to meet cold Polar air and thus develop cyclones or Lows.

Over Polar areas, especially in winter, the atmosphere also moves in large anticyclonic swirls, although these are transient. Tongues of this cold, dry, unstable Polar air sweep towards the Equator and cut beneath the lighter Tropical air.

The boundaries between Polar and Tropical air are the Polar Fronts, along which cyclonic swirls develop to spread rain and warmth to high latitudes. The boundary between the two Tropical air masses along the Equator is the Doldrums or Inter-tropical Front.

Pressure

WE HAVE NOW seen how air masses, of four main types, have vital parts to play in the type of weather experienced all over the world. The basic reason for these air masses moving in the way they do is *Pressure*. Pressure is the weight of air at any given point, and this weight varies, depending on *temperature* (warm air is lighter than cold); *humidity* (wet air is lighter than dry); and *density* (the thinner the lighter). Barometers react to such changes in weight by 'rising' when the pressure is High (usually giving good weather), and falling when Low (bad weather).

Pressure can also exert a force – high pressure forces air towards low pressure. This is the vital clue to all movement of air masses. Remember that (a) Arctic and sub-tropical areas are normally recognised as being high pressure zones, and (b) the temperate latitude area (about lat. 40°N–60°N) and the equatorial belt (about lat. 15°N–15°S) as being low pressure. These zones set a basic pattern of air movement, but a number of other factors can cause changes to it that are both predictable and unpredictable.

The most important of these factors is the geographical distribution of the major land masses, and the oceans in between. Their effect is to split up the basic pressure zones into *Pressure Cells*, around which the air moves in great swirls or cycles. The swirls around low pressure cells are *Cyclones*, those around high pressure cells, *Anticyclones*. The winds around cyclones are usually strong (frequently up to Storm Force 10), but within a large anticyclone the winds are light and variable over a large area, only increasing around the margins of the area as the pressure falls off – they seldom exceed Force 6 or 7.

The rotation of the earth imparts a circular movement to the air producing the swirls referred to above (the *Coriolis Effect*). Cyclonic circulation is counterclockwise in the Northern hemisphere, and clockwise in the Southern; anticyclonic circulation is the reverse.

The Sub-Tropical Anticyclones

The anticyclones of the world are the dominant factors in the creation of 'Weather'. In both Northern and Southern hemispheres, the sub-tropical ocean high pressure cells are permanent features. Situated between latitudes 10° and 40°N or S (approximately), they are the generating areas for Tropical Maritime air, as has been shown earlier, and they spread heat from the Tropics towards the Poles.

They pulse, of course – increasing in area and intensity from time to time, sometimes shrinking and weakening, sometimes breaking up into cells and sometimes re-forming. But always they are there, and around them blow the steadiest of all the world's winds – the Trade Winds and the Westerlies.

In the regular geographical confines of the North and South Atlantic Oceans this pattern is most stable. Throughout the year, and indeed throughout the centuries, the regular winds have moved the surface waters into the well-known ocean currents.

In the wider Pacific Ocean, some 6,000–8,000 miles of it, the sub-tropical anticyclones cannot maintain this east to west length in comparison to a 2,000 miles north–south width, and constantly break into cells. Locally these cells break the regular pattern of the surface winds but never long enough to destroy the Trade Wind regularity of direction and speed, so that the surface currents also never lose their regularity.

The Polar Anticyclones

In contrast to the tropical anticyclones the cells of high pressure which form over high latitudes are transient – but of no less importance.

They form slowly, often sluggishly, but when they do they last several weeks. Each becomes a mass of cold, dry, heavy air, 500 to 1,000 miles across – a new supply of the cleansing Polar Maritime air. For most of the year in the Northern hemisphere they form over Greenland or Scandinavia, but in winter are further south over the cold land masses of Canada and Russia.

Beneath these anticyclones the weather is consistently Fair or Fine, winds light and variable, with little or no precipitation of rain, snow, etc. Round their margins the wind becomes stronger (see Chapter 2). On the southern margins these stronger winds are NE to E in direction, and are blowing not far away from the equally strong SW to W winds on the northern margins of the sub-tropical oceanic anticyclones.

Thus, the Polar Front is not only a boundary between cold, dry, heavy Polar air and warm, moist, light Tropical air, but also a boundary between strong winds moving in opposite directions.

When these winds 'clash', violent swirls are formed, rotating counterclockwise in the Northern hemisphere and clockwise in the Southern, with pressure low at their centres and high to North and to South. These are the Temperate Cyclones which we usually call 'Lows' or 'Depressions'.

The Temperate Cyclones

These form quickly (a matter of hours), increase in area (up to 1,000 miles across within a few days), move quickly (up to 30 knots) and they move far.

Over the oceans their tracks are controlled by the anticyclones (see Fig. 14), until they reach land masses where mountain ranges slow them down or divert their path, sometimes halt them until their energy has been expended, mostly as rain.

In the Southern oceans, between latitudes 40°S and 60°S there is no land to impede their path, except the tip of South America. In these latitudes swirls form, increase in size and intensity and move eastwards in regular procession at speeds between 15 and 30 knots. Some fade away but many move round and round these Southern oceans, never really dying out. In the winter they move northwards to affect South Africa and SW Australia. New Zealand and Tasmania are affected by them throughout the year.

The Tropical cyclones

Between the tropical anticyclones to the north and south of the equator, lies the *Equatorial Doldrums*, zone of low pressure. Here two similar (Tropical Maritime), but locally different, air masses converge. At about 5° (say 300 miles/480 km.) to the north and south of the equator the Coriolis force becomes strong enough to give the necessary swirl-twist which gives birth to tropical cyclones – the *hurricanes* of the West Indies, the *typhoons* of the China Seas, with *Willy-Willys* off Western Australia, and the *cyclones* of the Indian Ocean.

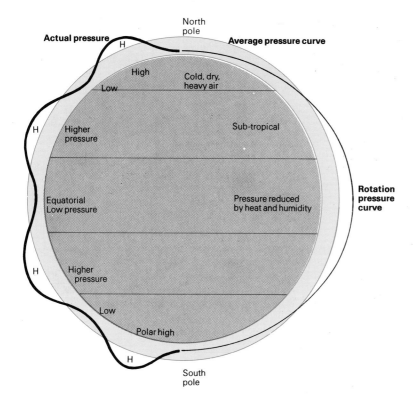

Fig. 13a Rotation of the Earth and pressure distribution. Rotation piles the atmosphere over the Tropics and pressure should accordingly be high. However, high temperatures and high humidity reduce

pressure around the equator. Conversely, low temperatures and low humidity increase pressure around the Poles. The result is the 'wavy' 'actual pressure' curve shown on the left of the diagram.

Frontal depressions

Whereas the tropical cyclone is small in area with only one type of air mass involved in its circulation, temperate cyclones involve the two, greatly differing, Polar and Tropical air masses and the dramatic demarcation line dividing them – the POLAR FRONT. As a result, the title 'temperate cyclone' is often replaced, in general usage, by *Frontal Depression*. The cyclonic swirl forms at the frontal boundary and distorts it into a wave. As the wave moves eastwards it becomes more and more pronounced, the swirl becoming larger in area and greater in vertical extent. The swirl may initially be 200 miles (320 km.) across but after several days can be up to 2,000 miles (3,200 km.), with the tropical air spiralling upwards to 40,000 or even 50,000 feet (15,000 m.).

Typical Mid-Atlantic Frontal Depression
Figs. 15 and 16 illustrate the development of a typical frontal depression. Such a depression would have the following characteristics:
Area: West to east 1,000 miles (1,600 km.) approx.
North to South 700 miles (1,120 km.) approx.
Speed: 25 knots (i.e. it would take about 40 hours to pass one particular point).

As will be seen from the diagrams, the depression is produced by the interaction of a warm front with a cold front.

The warm front

Slope of warm front: 1 in 125.
Altitude: Rising to 24,000 ft. (7,300 m.).
Cloud: An overcast layer of alto-stratus cloud about 300 miles (480 km.) ahead of centre of front.
Rain area: With a low nimbo-stratus cloud base

below 6,000 ft. (1,800 m.) the rain area extends about 125 miles (200 km.) ahead of the centre, giving a five hour period of rain.

Pressure: From 1,030 millibars ahead of the warm front, it falls to 980 mbs. at the centre. The average *Pressure Gradient* is 4 millibars per each 40 miles (64 km.). This means that isobars drawn at 4 mbs. intervals will be about 40 miles apart on the weather map.

Wind speed for warm front

A pressure gradient of 4 mbs./40 miles (64 km.) will give a maximum wind speed at the surface of 40 knots, Gale Force 8, which is average for a depression of this depth.

Barometer fall for warm front

The average rate of fall of the barometer for such a depression is $2\frac{1}{2}$ millibars per hour. For the first 5 hours the actual rate will be low (1 mb./hour) though enough to give adequate warning that the Low is approaching. Over the next 5 to 10 hours the rate of fall increases and the wind moves from Force 4 to Force 6 and stronger. When the rate is 3 mbs./hour, a full Gale Force 8 can be expected, though it will probably take a further 3–4 hours before the wind reaches that strength. A shallow, fast-moving depression will reach peak wind speed considerably earlier than a slow-moving depression. Either way there is time to take precautionary action both ashore and afloat in readiness for the gale – which, in fact, may not materialise if any one of a number of factors change. But best be prepared!

*Fig. 13*b. A *Temperate cyclones (Lows) form here.*

B *Tropical cyclones (hurricanes) form here.*

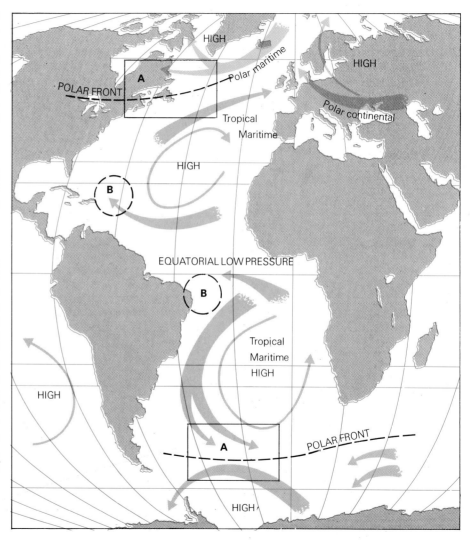

The cold front

Slope of cold front: 1 in 50.

Speed of cold front: Greater than the speed of the warm front and about the same as the NW wind behind the cold front (in this example about 40 knots).

Note: There is no fall of pressure to herald the approach of a cold front, as there is ahead of a warm front.

Pressure: The rapid increase in the vertical extent of the heavy Polar air behind a cold front at an average slope of 1 in 50 is reflected in the rapid initial rise of the pressure. This steep rise-rate then falls off. In the first five hours, it can be up to 30 mbs., a rate of 6 mbs./hour; in the second five hours the rate will have fallen to 2 mbs./hour.

Wind speed for cold front

At the passage of a cold front, the wind veers very suddenly from SW to NW and the strength increases to Force 8 or even Force 9 with vicious suddenness. Because this cold Polar air is very unstable, the degree of gustiness for both speed and direction is very high and this is greatly accentuated in the squalls beneath the lines of cumulo-nimbus cloud.

The north-westerly gales can be expected to last for about four hours by which time the rate of rise of pressure should decrease to about 2 mbs./hour and the wind should moderate down to Force 6. After ten hours the rate of fall should decrease to about 1 mb./hour and the wind moderate to Force 4.

If this expected rate of rise does *not* continue for these estimated times but falls off rapidly, with the wind moderating more quickly and possibly backing from NW to W, BEWARE. Such quick improvement is too good to be true. It is an almost certain indication of the formation of a secondary 'wave' depression, away to the south-west forming on the trailing cold front.

Secondary 'Wave' Depressions

These form on cold fronts and move very rapidly, at over 45 knots, often covering over 300 miles (480 km.) in six hours. They are preceded by overcast conditions, low cloud and rain, and some by gale force southerly winds, just as for a normal warm front in a normal depression. They are followed by the passage of the same cold front for the second time.

It is possible for a series of waves such as these to run along a cold front, the general line of which can remain stationary for several days. The area along that general line for a width of 100–200 miles can experience continuously unsettled weather, with no marked clearances between periods of overcast and rain interspersed with short cloudy periods and showers. The wind will alternately back to southerly, then veer to north-westerly, and the temperature and humidity will just as abruptly rise and fall, as Polar and Tropical air alternate over any one locality.

Fig. 14 (right). How a depression moves. The cyclone is like a small cog-wheel moving between two larger, anticyclonic cogs. In the Northern hemisphere the movement of depressions (Lows) is controlled by the anticyclones, which are pools of cold, heavy air. 1. The cyclone forms at A and usually moves to B where it becomes larger. It then moves northwards to C as it links with anticyclone Y. If Z remains dominant the Low will move to D. 2. If anticyclone X is strong, Low A may move westwards. This is how depressions affect the north eastern parts of N. America.

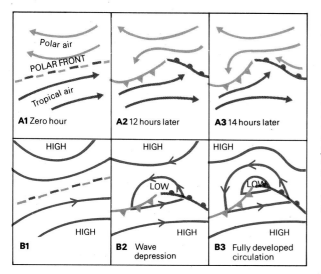

Fig. 15 (top left). A Air flow; B Isobars and wind direction. On the Polar side of the Polar Front is Polar air moving westwards very rapidly; on the other side is Tropical air moving eastwards equally rapidly. Friction along the boundary causes a cyclonic swirl (counterclockwise) to form as shown. A depression or Low has been created. In the Southern hemisphere the picture is a mirror image and directions are reversed.

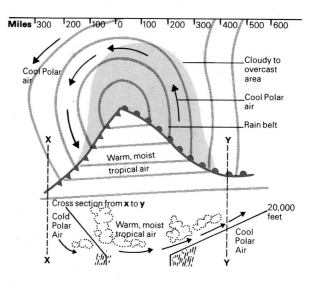

Fig. 16 (top right). Pressure pattern of a typical depression. As the swirl moves west to east, we experience weather from fine to cloudy, to rain, to cloudy and back to fine again.

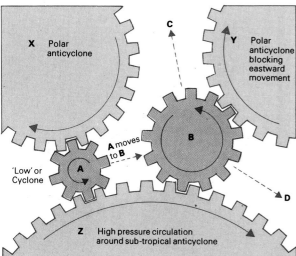

Fig. 17 (bottom right). Pressure pattern of a typical depression. How the barometer rises and falls.

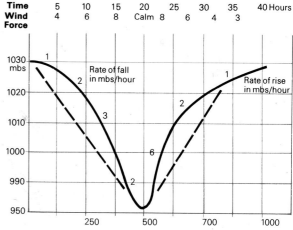

Chapter 6

WEATHER ON LAND is characterised by extremes —
hotter by day and cooler by night, very much
hotter in summer and very much colder in winter.
Central Asia, for example, becomes covered in
winter by a pool of cold, dry, heavy Continental air.
This high pressure complex pushes winds outwards,
giving spells of intense cold over NE Europe, and
producing the NE Monsoon over SE Asia.

In summer pressure becomes low in the intense
heat, resulting in great waves of rain-laden Tropical

Weather on land

Maritime air being sucked in, e.g. the SW Monsoon. This monsoon effect is occasionally experienced on a smaller scale over North America and Australia.

Altitude reduces temperatures. Mountain ranges impede the flow of cloud and rain-bearing Tropical Maritime air over and into the larger continental masses.

The further from the sea, the drier and finer becomes the weather, and the sparser becomes the natural vegetation.

Majestic view over the Black Hills in South Dakota.

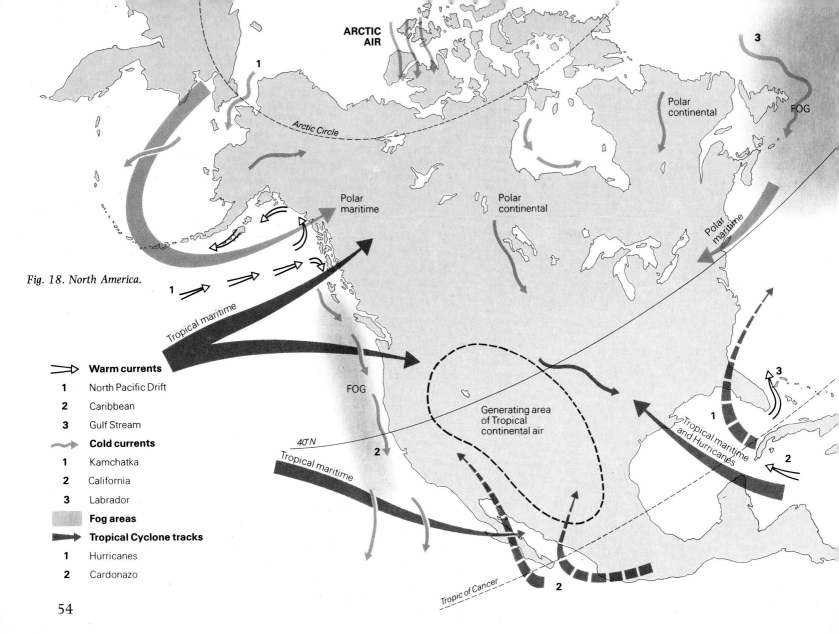

Fig. 18. North America.

ARCTIC
AIR

Arctic Circle

Polar
continental

FOG

Polar
continental

Polar
maritime

Polar
maritime

1

1

Tropical maritime

Warm currents

1   North Pacific Drift

2   Caribbean

3   Gulf Stream

Cold currents

1   Kamchatka

2   California

3   Labrador

Fog areas

Tropical Cyclone tracks

1   Hurricanes

2   Cardonazo

FOG

Generating area
of Tropical
continental air

40° N

Tropical maritime

2

Tropical maritime
and Hurricanes

3

1

2

Tropic of Cancer

2

54

The North American continent

The vast size of this land mass – extending from the Tropics to the Arctic – exposes it to all types of weather, as differing air masses swirl over it. The continuous high wall of the Rocky Mountain complex stands up into the atmosphere like a dam across a river, to impede the movement of Tropical Maritime air from the Pacific, and the spread of warmth and rainfall into the interior. To a lesser extent the Appalachian system in the east prevents warmth and rain being carried westwards by depressions formed over the Atlantic.

The interior of the continent, however, is open to swirls of Tropical Maritime air coming in from the south-east, either as Temperate (relatively gentle) or as Tropical (viciously active) cyclones called *hurricanes*.

In the north, there is no protection against Polar air, or the even colder Arctic air, sweeping down off the frozen wastes of the Arctic, to spread clear, dry, but bitterly cold weather southwards, in winter even as far as the Gulf States, when ice-bound conditions can exist as far south as St. Louis.

Air masses and related weather

Tropical Maritime air from the Pacific. The temperate latitude cyclones (Lows) of the north Pacific bring pulses of cloudy-to-overcast weather, with periods of moderate to heavy rainfall, to the western seaboard from Alaska to, in winter, California. These swirls do not easily cross the mountains, so cloud and rainfall tend to be concentrated between mountain and coast. The interior, in the *rain-shadow*, is clear and dry. When they do, however, they bring westerly winds over the Rockies. The air which rises over the mountains forms cloud and deposits rain. Then, as it descends on the eastern side of the range, it becomes the warm, very drying wind known as the *Chinook*.

Tropical Maritime air from the Atlantic. Temperate latitude depressions forming over the western North Atlantic, can swirl north-west or west. Their subsequent path can be affected by anti-cyclones developing in the area. For example, a strong High over Newfoundland could steer a Low developing over Bermuda on a track over Cape Canaveral towards the Great Lakes.

Tropical Cyclones, geographically small, but great in energy, can swirl over the Southern States deep into the heart of the continent before curving away to the north or north-east.

Polar air. Polar Maritime air is sucked into almost all temperate cyclones so that clear, cold but showery weather, often with squalls and thunder, rapidly replaces, as the 'cold-front' moves in, the warm, moist tropical air ahead of and in the warm sector of the cyclone.

Polar Continental air is not so much sucked in behind depressions as pushed outwards by vigorous anti-cyclones or Highs, which are vast areas of cold, dry, heavy air. Such Highs over the northern islands can push Polar or Arctic air far, far south, especially if further High cells over the western prairies assist this movement.

Arctic air is so named as it is colder and drier than Polar air; it can, in winter, generate anti-cyclones to the north of the continent, and intensely cold waves can spread far south into the interior of Canada and the United States.

Tropical Continental air. The vast dry areas of Texas, Mexico, New Mexico and Arizona are generating areas for hot, dry Tropical Continental air.

Fig. 19. Australia.

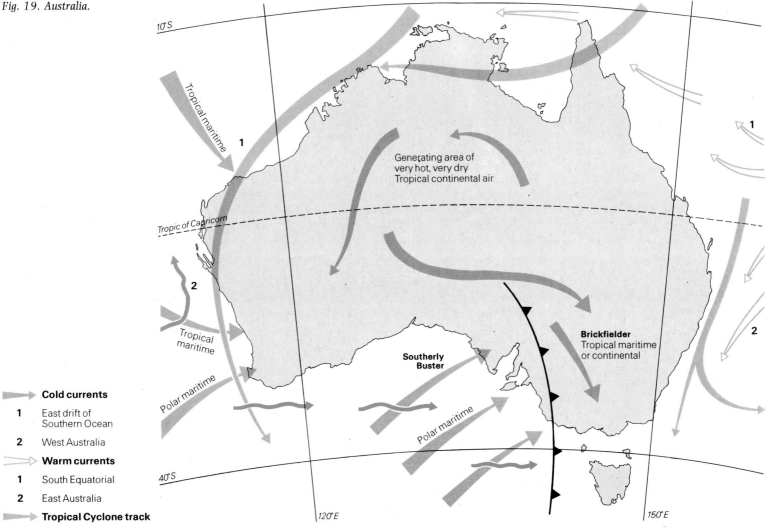

10°S

Tropical maritime

1

Generating area of
very hot, very dry
Tropical continental air

Tropic of Capricorn

2

Tropical
maritime

Southerly
Buster

Brickfielder
Tropical maritime
or continental

Polar maritime

Polar maritime

1

2

Cold currents

1 East drift of
Southern Ocean

2 West Australia

▷ **Warm currents**

1 South Equatorial

2 East Australia

Tropical Cyclone track

40°S

120°E

150°E

The weather of Australia

This continent lies across the Tropic of Cancer from latitude 10°S to latitude 50°S (some 1,800 nautical miles) and extends from longitude 115°E to 155°E (some 2,000 nautical miles). Its weather is determined by the air mass as modified by local geography, and by the ocean currents which modify the air masses passing over them.

Air masses

Tropical Maritime air. In summer the continent is always covered by Tropical Maritime air, bringing cloudy weather and rain off the surrounding tropical seas. The maritime influence decreases inland, especially in the east, where the mountain ranges from Cape York to Cape Howe form an effective barrier.

Tropical Continental air. This continent is large enough (some 3 million square miles) for the interior to be a generating area for Tropical Continental air. It is possible for a vast pool of air to come to rest over the interior, to become dry and very hot, but still heavy enough for the surface pressure to be greater than nearer the sea. High cells form from which hot dry winds blow anti-cyclonically outwards (counterclockwise in direction) but may then be sucked into the cyclonic (clockwise) circulation of depressions passing to the south. Such a situation generates the north to north-westerly *Brickfielder* winds.

Polar Maritime air. A steady sequence of temperate latitude depressions encircles the world between latitude 40°S and latitude 50°S. This belt is known as *The Roaring Forties* because of the frequency and violence of the westerly gales which are its main feature. Behind each depression comes a wave of heavy, cold, unstable Polar Maritime air, surging northwards off the Antarctic ice.

In summer these depressions and waves of Polar Maritime air only occasionally push further north than Tasmania, but during the winter months they regularly affect the southern states to give them the classic 'Mediterranean' type of climate and weather.

Ocean currents

The east drift of the ocean surface water to the south of Australia is under the influence of the westerly winds of the Roaring Forties. This produces a steady flow of water which is markedly colder than the corresponding area of the North Atlantic ocean in latitude 40°–50°N. This surface water is fed from the melting ice whose limits are relatively close – drift ice at 50°–55°S and pack ice at 55°–60°S, only 500–1,000 miles from the southern shores of Australia. This drift runs northwards up the west coast, taking its cooling effect into the tropics.

In the east and to the north, the westward drift of the equatorial surface water across the Pacific turns south along the Great Barrier Reef to keep eastern coastal waters warmer than average.

Tropical cyclones

These small but intensely vigorous swirls of Tropical Maritime air generate over the Timor Sea and Sea of Arafura. Typical subsequent tracks are shown on the map opposite. These cyclones cannot penetrate far inland as a supply of moisture is needed to maintain their energy. Until they lose this energy, however, they are the most destructive of weather phenomena. The most recent example to hit Australia was the devastating cyclone which struck Darwin on Christmas Day, 1974.

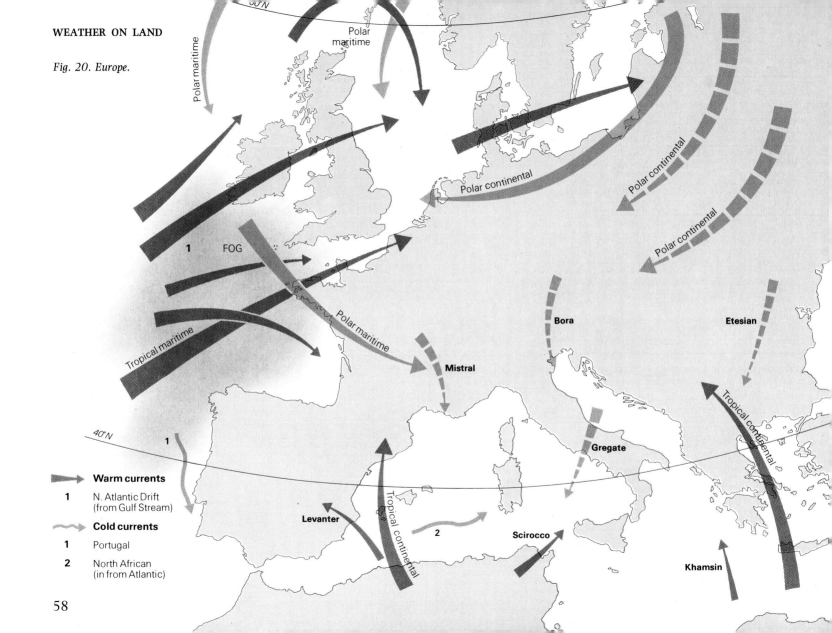

Fig. 20. Europe.

Polar maritime

Polar maritime

Polar continental

Polar continental

Polar continental

1

FOG

Polar continental

Polar maritime

Tropical maritime

Bora

Etesian

Mistral

Gregate

Tropical continental

40°N

1

Levanter

Tropical continental

2

Scirocco

Warm currents

1   N. Atlantic Drift
(from Gulf Stream)

Cold currents

1   Portugal

2   North African
(in from Atlantic)

Khamsin

The weather of Europe

Europe lies mainly between latitude 40° and 60°N with Scandinavia and Russia extending to 70°N. The weather is therefore both extremely varied and variable, since every permutation of both major types of air mass can and do affect all European countries. Local weather is of course greatly affected by local geography.

Air masses

Polar Maritime air. High pressure over Greenland combined with high pressure over South-east Europe can bring great waves of cold, relatively dry, clear Polar Maritime air sweeping over all Europe. There will be some local showers over both land and sea by day (but not over land at night), caused by convection, but otherwise the weather will be similar over the area covered by the air mass.

Polar Continental air. High pressure over Scandinavia will bring waves of Polar air sweeping southwards and then westwards over most of Europe. In winter these winds will be bitterly cold and dry; the type which brings the coldest spells of weather to all Northern Europe. The skies may be clear but the short hours of daylight, however sunny they may be, cannot balance the long chill hours of clear nights. So long as this air stream persists, temperatures will remain below freezing. In summer the reverse occurs.

Tropical Maritime air. The generating area of Tropical Maritime air is the *Azores High* over the North Atlantic, and this air invades Europe not so much as waves but in the swirls of Lows or depressions (temperate latitude cyclones). The Tropical Maritime air then moves in along the cyclone tracks which, in turn, are modified by the continental geography.

In summer the tracks are further north; in winter further south and into the Mediterranean area, bringing with them the traditional winter rainfall of the Mediterranean climate.

Local winds are the products of the paths of depressions combined with local geography, especially around the Mediterranean. e.g.:

MISTRAL and BORA: cold, strong northerly winds, increased locally by the downhill, katabatic effect of Polar air behind the cold fronts of depressions moving east along the north European lowlands.

LEVANTER is the warm, moist easterly wind which blows across Gibraltar. It is created by cells of high pressure which form over Spain and France.

LLEVANTADES and GREGALE are local names given to strong north to north-east winds associated with depressions moving east along the central Mediterranean, and again the air is of Polar origin, unstable with squally showers and thunderstorms. Ahead of these depressions occur:

SCIROCCO and KHAMSIN, the hot, dry, dusty winds of Tropical Continental air which blow north off the Sahara. They can become moist if they spread across the Mediterranean to reach the European coast. These winds are quite common.

Ocean currents

The North Atlantic Drift spreads the warmth of the Gulf Stream in a wide arc from Iceland and the North Cape to Biscay, to keep all ports in its path ice-free, and to ameliorate the cold impact of Polar Maritime air masses.

Nevertheless, sea temperatures from Ireland to Portugal are low enough to cool Tropical Maritime air streams to the dew-point, so producing extensive fog banks.

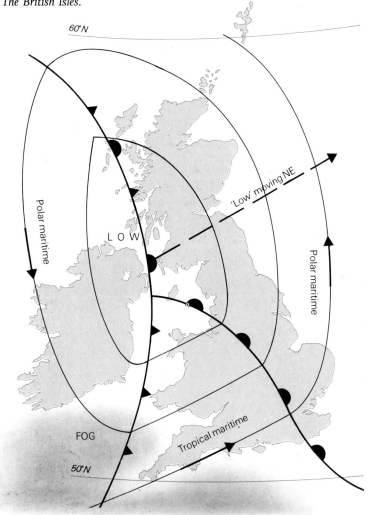

Fig. 21. The British Isles.

Island weather —
The British Isles

The normal weather pattern is of an endless succession of depressions (Lows), all giving similar weather, but no two ever exactly alike, nor ever following exactly the same track at exactly the same speed.

Returning Polar Maritime air. Ahead of a depression, returning Polar Maritime air gives moderate but freshening S–SE winds. Fair weather at first, becoming cloudy, then rain spreading from the west.

Tropical Maritime air of the warm sector brings on wind-veer to south-west. The overcast breaks up to cloudy or fair, rain gives way to drizzle, but fog banks at sea, or hill fog along the coasts especially, are the common weather hazards. Polar Maritime air sweeps in from the west or north-west behind the cold front. Squalls, thunderstorms and hail accompany this wind-veer, but visibility improves to excellent. Gradually this air stream moderates, the showers die away as the depression moves away. The weather improves to fair, even to fine, to remain like that until the next depression brings more cloud and rain.

Long fine spells occur over the British Isles only so long as a Polar Continental air stream persists, i.e. with an anti-cyclone over Scandinavia.

In winter Polar Continental air brings cold dry weather to all U.K. Should any vigorous depression approach from the Atlantic, the swirl of moist, warm Tropical Maritime air will deposit snow for a few hours, but the weather will quickly become cold and fine again.

In summer dry warm east to south-east winds can persist for weeks and prevent any depressions from bringing rain and bad weather.

Island weather – New Zealand

The weather of New Zealand is a succession of Lows very similar to that of the British Isles, with the same resulting air masses and weather patterns. The main difference is that the circulation of winds around these Lows is clockwise, and that the Lows move from NW to SE. New Zealand, latitude 35°S to 45°S, is closer to the Equator than Britain (50°N to 60°N), but to the south, the limits of drift and pack ice are closer than equivalent areas are to Britain.

Tropical Maritime air is somewhat warmer and carries more moisture. Average temperatures are 58°–66°F (compared with 40°–60°F of U.K.), and rainfall exceeds 50 inches over a greater proportion of the Islands.

Polar Maritime air is rather cooler than might be expected. The ocean crossing from the ice limit is between 500 and 1,000 miles (20 to 50 hours for air moving at 25 knots, force 6) and the temperature of the surface water is normally very little above freezing point.

There are no large land masses to generate any Continental type of air, either Polar or Tropical, so New Zealand rarely experiences long periods of settled weather comparable to Britain's occasional winter 'freeze-ups' or summer heatwave/droughts.

Geography. The SW–NE lie of New Zealand and of its mountains is across the path of depressions and of the rain-bearing Tropical Maritime air from the north-west. Hence rainfall is concentrated on the west, and the eastern lowlands have more sunshine, less rain and less unsettled weather.

In general New Zealand is that bit milder and wetter than Britain, and rarely experiences extremes of winter cold or of summer heat-wave.

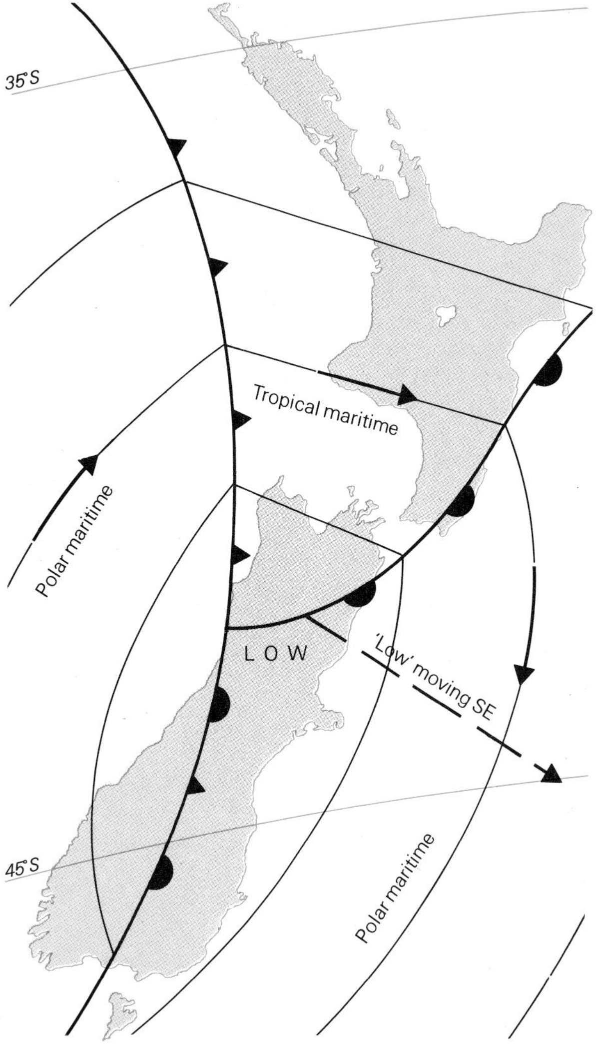

Fig. 22. New Zealand.

61

Land and sea breezes

Land surfaces, especially of bare rock, warm up rapidly by day to give maximum temperatures about one hour after mid-day. At night they cool, unless protected by a cover of cloud, until about an hour after sunrise this radiation cooling is halted by the rising sun. This day–night *diurnal* temperature range can be up to 86°F (30°C) in sub-tropical lands, reducing to 50°–59°F (10°–15°C) in temperate latitudes.

Sea surfaces, in marked contrast, maintain steady temperatures. Diurnal range rarely exceeds 36°F (2°C) even in shallow waters, and seasonal ranges are between 41° and 50°F (5° and 10°C). Both depend on ocean currents such as the Gulf Stream rather than on the balance between day-time insolation and night-time radiation.

Sea-breezes therefore set in during mid-forenoon as soon as the land is appreciably, say 5–10°C, warmer than the off-shore sea surface, and persist until mid or late afternoon. Their force depends on the temperature differences which in turn reflect local topography. They can reach 15–20 knots, force 4–5, extend up to 1,000 ft. (300 m.) and inland for 5–10 miles (8–16 km.), but much further than this along low-lying arid coasts. These cool, moist but salt-laden breezes have a marked effect on coastal vegetation.

Land-breezes set in by night but it is usually near dawn before the temperature differences are sufficiently great to trigger them off. Their vertical extent is usually less than 500 ft. (150 m.) but they extend further out to sea because the sea surface friction is less than that over land.

Land and sea breezes are most noticeable when the weather is otherwise calm but, except in cloudy or overcast weather the temperature differences will be present to cause these localised pressure differences and the resultant breeze. Hence the usual winds dictated by the prevailing isobaric pattern will be modified wherever land and water are together, whether that be along oceanic shores or around much smaller areas of water-meadow or marsh.

Mountain breezes
By day the higher slopes of hills and mountains especially if facing eastwards, are warmed by the rising sun earlier than the shaded valleys. Locally the warmer, expanded air rises as *thermals* to be replaced by wind blowing uphill from the valleys. Called *anabatic* winds they are more common than was once supposed, but are experienced by all mountaineering expeditions and by all aviators. Usually relatively gentle, they can be locally and temporarily very strong and can be superimposed on the general pressure-pattern of winds.

By night the reverse process occurs, the air at altitude becoming very cold very quickly after sun-down. Gravity then takes over and the cold air streams down into the valleys like torrents of water. On a small scale it can create frost hollows: on a large scale it creates the notorious fierce winds of Norway and of Greenland. In Antarctica these *katabatic* winds as they are called have been recorded up to 80 knots.

Mountain coasts
Along mountainous coasts day-time sea breezes and anabatic winds can reinforce each other especially where the mountain ranges are parallel to the coast. By night (or, more correctly, nearer dawn), land-breezes and katabatic winds also

reinforce each other, but this time most markedly so when valleys, e.g. fjords, lead down to the sea. The total result, as often experienced along Mediterranean coasts, can be an off-shore wind of gale force sweeping from a valley system some 20–30 miles (32–48 km.) out to sea before its force is absorbed.

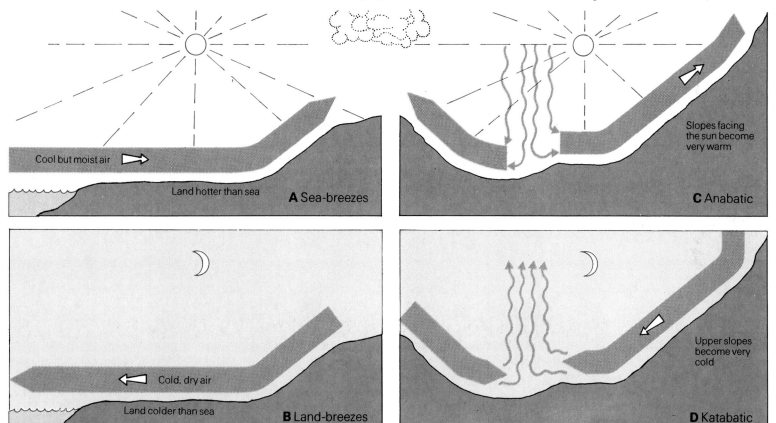

Fig. 23. Local winds are the result of local differences in heating or cooling. A Sea breeze by day. B Land breeze by night. C Mountain winds: warm air rises above the sunny slopes and creates up-currents. D Mountain winds: cold air sinks down into the valleys.

Mountainous coasts can experience coastal breezes and mountain winds at the same time.

Cool but moist air

Land hotter than sea

A Sea-breezes

Slopes facing the sun become very warm

C Anabatic

Cold, dry air

Land colder than sea

B Land-breezes

Upper slopes become very cold

D Katabatic

Chapter 7 Weather at sea

THE SAILOR'S HABITAT is not so much the sea as the surface of the sea. The wind acts on this surface and produces *waves*, undulations in the surface on which a vessel is floating. These waves are in harmony with the speed or force of the wind, its direction, its gusti- ness for both speed and direction, the time for which it is blowing and, lastly, the distance over which it is blowing. Waves of the most uniform character are found where the wind blows at a relatively steady strength for long periods of time over long distances.

Sea and swell

Sea waves (or more simply, *Sea*), is the name given to the waves created by the wind actually blowing over an area, which is then called the *Generating Area* for those waves.

Swell waves (or simply, *Swell*): once waves have been created they persist, running ever onwards, relentlessly keeping their direction, their length, their speed (and therefore their period). The steady reduction in their height, as their energy is very slowly dissipated, is the only property to change over the open sea. Eventually, of course, they will come into water shallow enough for the sea-bed to drag and hold back the deeper part of the wave motion while the surface layer runs on unimpeded. As a result the wave *breaks*; the wave motion is translated into lateral motion of the water of the upper part of the wave.

Waves will thus keep 'running' even after the wind that produced them has died away – or after they have moved away from the area in which they were generated. They are then known as *Swell waves*, as distinct from *Sea waves* (see above).

Understandably it is very rare for there to be no swell. If there is swell present then the sea waves created by a *new* generating wind will be different from those which would be created were there no swell present. This is an important factor to bear in mind when attempting to forecast conditions in the next twelve to twenty-four hours.

Pressure, tides and storm surges

In addition to causing the winds to blow, pressure affects the height of sea level. When the pressure is high, the heavy air forces the sea level down; when it is low the level rises accordingly.

When the barometer shows a High Pressure of 1,030 mbs, both High tides *and* Low tides will be *lower* than predicted for normal conditions by 1 ft. (30.5 cm). If the weather is fine, watch for this.

With Low Pressure the reverse is true. A deep depression of 980 mbs will *lift* both High and Low waters by 1 foot (30.5 cm.). It will create a 'surge' beneath itself, moving with it and behaving very much as the normal upward surge of the tide moves under the influence of the moon. Deep depressions moving rapidly over confined waters can create storm surges of up to 8 ft. (2.5 m.) or more, because to the pressure effect is added the effect of the strong winds inevitably associated with them. This phenomenon has caused major flooding disasters along the southern margins of the North Sea throughout history, the Low Countries suffering most.

The state of the sea

The *State of the Sea* is the last factor making up the Sea-farer's environment. Above him are the wind and weather – the clouds, the rain and the hail; around him the moisture and the dust which restrict his visibility; beneath him the waves – the sea and the swell. As the weather is the creation of the wind, so are the waves. As all weather factors depend on the direction and speed of the wind, so do the waves in all their properties of height, length and speed. The waves lie across the wind; they move with it; they take their energy from it.

For any given wind speed, the waves created will gradually increase in height, length and speed until they reach maximum values for that wind speed. Thus, it takes both time and distance for waves to reach their maximum development. This distance is called the *Fetch*.

Sea Scales

The photographs in this series illustrate the precise meanings given the words used in weather reports and forecasts to describe the *State of the Sea*. The words also have a number in the Sea Scale devised by Admiral Douglas.

Here the surface of the sea is glassy, almost like a mirror. Yet the surface is not completely without motion as *swell* waves with clearly visible troughs and crests can be seen.

Douglas Sea Scale: 0
Wind: Beaufort Scale Force 0 – 'Calm'.
Speed: Less than 1 knot.

Calm

The sea is still *Calm* but the wind of mean speed 2 kts. is enough to create ripples with the appearance of scales, but without any crests or any foam.

In this photograph, too, can just be discerned the crests and troughs of a family of *swell* waves.

Douglas Sea Scale: 0
Wind: Beaufort Scale Force 1 – 'Light Air'.
Speed: 1–3 knots.

Smooth

This wind, of mean speed 5 kts. is now strong enough to create wavelets, only small (a few inches in height) but with pronounced crests, although these are too small to break.

However, these waves make it difficult to see any *swell* waves which might be present although such waves would transmit their motion to small boats to make them pitch or roll.

Douglas Sea Scale: 1
Wind: Beaufort Scale Force 2 – 'Light Breeze'.
Speed: 4–6 knots.

Slight

The wind Force 3 is strong enough in two to three hours to produce wavelets up to a mean height of $1\frac{1}{2}$ ft. (50 cms.). Here and there the crests break to give a few small 'white horses'.

The crests and troughs are beginning to settle into a pattern sufficient for dinghy sailors to be aware of them.

Small cruising yachts sailing into the wind – and the waves – might put up a little spray.

Douglas Sea Scale: 2
Wind: Beaufort Scale Force 3 – 'Gentle Breeze'.
Speed: 7–10 knots.

Moderate

The moderate sea produced by the moderate breeze has longer, more pronounced crests some 2–3 ft. (60–100 cm.) high, and some of them breaking into 'white horses'.

These conditions are ideal for dinghy sailing in the open sea though the pressure of racing will probably produce a capsize or two and the presence of a safety-boat is therefore essential.

Cruising in a Force 4 is ideal; but racing really requires a stronger wind.

Douglas Sea Scale: 3
Wind: Beaufort Force 4 – 'Moderate Breeze'.
Speed: 11–16 knots.

Moderate

A further increase in wind speed to Beaufort Force 5 serves to accentuate the basic character of a moderate sea, namely wave height 4–5 ft. (1–1.5 m.) but with many crests breaking into 'white horses'.

Dinghy sailing in this state of sea becomes 'exciting', and frequent capsizing must be expected. The cruising enthusiast will want to reduce sail and lay-off the wind a little; the racing enthusiast will tighten his sheets and push her along.

Douglas Sea Scale: 3
Wind: Beaufort Force 5 – 'Fresh Breeze'.
Speed: 17–21 knots.

Rough

By now the waves are large – between 5 and 9 ft. (2–3 m.) with white foam crests everywhere. The wave speed will be of the order of 15–20 knots, and the wave length between 100 and 200 ft. (300–600 m.).

It takes about 12 hours, with the wind acting on these waves for a distance of at least 100 miles (the fetch), for this degree of development to be reached.

In these conditions a yacht 'hove-to' will drift to leeward at about 1 knot. 'Reaching' across the wind the *'leeway'* may well be over 10°.

Douglas Sea Scale: 4
Wind: Beaufort Scale Force 6 – 'Strong Breeze'.
Speed: 22–27 knots.

Very rough

Up to 1958 Beaufort Wind Force 7 used to be called a Moderate Gale but although the term *Gale* is used gale warnings are not issued unless the mean wind speed is expected to reach Force 8.

This photograph shows the state of the sea created by a Force 7 blowing for some 12 hours over a fetch of a 100 miles.

The sea is heaped up and the white foam from the breaking waves is blown in streaks along the direction of the wind.

Douglas Sea Scale: 5
Wind: Beaufort Scale 7 or Near Gale.
Speed: 28–33 knots.

High

After some 12 hours Gale Force winds blowing over a fetch of 100 miles will create waves up to 20–25 ft. (8 m.) in height moving at 25 knots, length 300–400 ft. (90–120 m.), period 6 or 7 seconds.

The tops of these waves are blown off in well-marked streaks of spindrift.

Where the depth of water is less than the wave-length i.e. 150 ft. (45 m.) friction-drag on the sea bed causes waves to become steeper and to break. Hence shoal areas are marked by broken, confused water.

Douglas Sea Scale: 6
Wind: Beaufort Force 8 – 'Gale'
Speed: 34–40 knots.

Very high

The waves are indeed very high at 25–35 ft. (7–10 m.) and if crests get in phase, freak heights of 50–60 ft. (15–18 m.) must be expected. Crests of waves really break. Wave motion of around 30 knots is translated into lateral movement of a vast bulk of water, which, breaking over a vessel, can create intensely high air pressure within the hull, enough to blow outwards any areas of weaker structure such as hatch covers or dead-lights.

Douglas Sea Scale: 7
Wind: Beaufort Scale 9 – 'Strong Gale'.
Speed: 41–47 knots.

Sea and swell waves

This photograph was taken off the Bahamas, looking Southwestwards, and shows *swell* waves, generated by the very high winds of a Hurricane centred off the West Indies, running out of the generating area and now invading this area from the South. The local Wind is Westerly Force 4–5 and, in harmony, the local State of Sea is 'Moderate', Douglas Scale 3, with Wave Height 2–4 ft. and Speed 5–10 knots.

The short wave length suggests that this hurricane is not yet fully developed. The low wave height suggests that it is some distance away, probably about 900 miles (1,400 km.) or some 30 hours for waves moving at 30 knots.

A waterspout

This photograph, taken in the Western Mediterranean in November, shows the vigour of the swirling turbulence beneath all big Cumulo-Nimbus clouds, even if they are not always strong enough to generate whirlwinds.

Sub-tropical waters retain their heat through the Autumn and Winter sufficiently to generate violent convective turbulence in an invading stream of cold, unstable, Polar air.

The same whirling turbulence is also possible over land, when whirlwinds, tornadoes or 'twisters' are formed. The photograph on the left is of a waterspout observed at San Feliu De Guixols, Spain, in September, 1965.

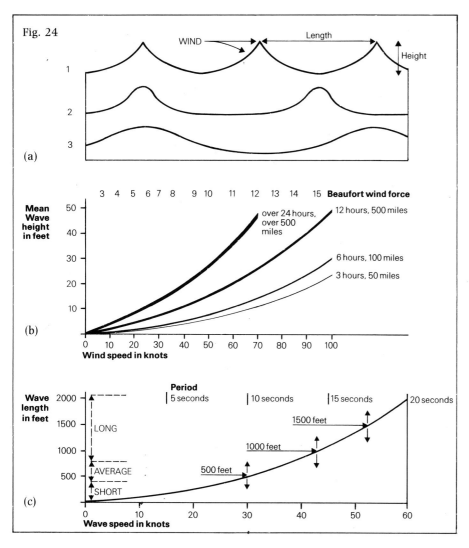

Fig. 24

(a)

WIND

Length

Height

(b)

3 4 5 6 7 8   9 10   11   12   13 14   15 **Beaufort wind force**

Mean
Wave
height
in feet

over 24 hours,
over 500
miles

12 hours, 500 miles

6 hours, 100 miles

3 hours, 50 miles

Wind speed in knots

(c)

Wave
length
in feet

Period
5 seconds   10 seconds   15 seconds   20 seconds

LONG

AVERAGE

SHORT

1500 feet

1000 feet

500 feet

Wave speed in knots

Wave developments

When waves first begin to develop under the influence of a light wind of Force 2 or 3 they have what is known as a *Cycloidal* shape (see Fig. 24(a)). These waves are of relatively short length but their steepness (their height in relation to their length) is increasing so that occasionally the limit of steepness is reached and the wave crest *breaks* as a 'white horse'. As soon as this occurs the steepness ratio is instantly reduced and the wave regains its equilibrium and shape. With any given wind force, wave height will increase relatively quickly over the first 2–3 hours, until the waves are about 75% of the mean heights listed in the Beaufort Scale, but after that the increase is slow. More than 24 hours of time and over 500 miles of Fetch is needed for waves to attain their full development in height, length and speed.

Moderate to fresh winds
Under the increasing energy of winds of Force 4, 5 or 6 the shape of typical waves becomes *Trochoidal*, (see Fig. 24(a)). These are mature, well-developed, deep-sea waves illustrated on pages 68–69. Provided the wind is blowing this is the characteristic pattern. In deep water, so long as these are the only waves, the chances of waves running together to form abnormally high crests or low troughs are relatively small, and no great problems of seamanship should therefore occur.

Strong winds
Increase of wind strength up to Force 9 or even Force 10 does not greatly change the trochoidal shape of the waves except that the crests, exposed to the full force of the wind, are frequently destroyed, flattened

as spume, or even blown off as driving spray. What does happen is that in addition to the wave height increasing to its maximum, so do wave-length and speed also increase to their maximum values.

Swell

When the wind ceases to blow over waves, their surface shape changes to become *sinusoidal* (see Fig. 24(a)). The surface becomes smoother, the crests more even. It is the old wave, the swell, running on, its speed and length unchanged but its height slowly decreasing until it eventually reaches shallow water and ends life as surf on a beach.

Swell is described by its length and height:

LENGTH		HEIGHT	
Short:	up to 300 ft. (90 m.)	*Low:*	up to 6 ft. (1.8 m.)
Average:	300–600 ft. (90–180 m.)	*Moderate:*	6–12 ft. (1.8–3.6 m.)
Long:	over 600 ft.	*Heavy:*	over 12 ft.

Wave height

Fig. 24(b) indicates how the Mean Wave heights can be developed in deep water in periods from 3 hours up, to over 24 hours, and with the Fetch increasing from under 50 miles (80 km.) to over 500 miles (800 km.). This diagram is simplified and so is not wholly accurate; however it gives an acceptable approximation. The upper curve shows the figures listed in the Beaufort Scale.

Wave length and speed

Fig. 24(c) shows the relationship between wave-length, speed and period. Waves 500 ft. (150 m.) long move at 30 knots, and the period is 10 seconds; 1,000 ft. (300 m.) long at 40 knots with a 15 second period and so on. Conversely, if the period can be measured – as it very easily can, either by counting the seconds mentally, or more accurately using a watch – the speed and length can then be assessed.

In relatively sheltered waters the first sign of swell is usually given by vessels at anchor. They begin to roll, almost imperceptibly at first, but their masts act like the pointers on metronomes, exaggerating the slight movement of the hull into a wide arc at the mast-head.

As vessels roll slowly in a swell which is otherwise invisible, timing of that rolling will immediately give the length and speed of that swell. Speed in knots is 3 × period in seconds. From this can be deduced the strength of the wind which generated the swell. Swell is, therefore, an advance gale warning – the classic mariners' warning of the existence of tropical cyclones.

Freak waves

The maximum theoretical height of a wave is 1.4 × its mean height, but much higher freak waves are possible, and should be prepared for. Waves of 'freak' heights have been experienced and survived even when they have broken over vessels. It is always possible for waves to be travelling at slightly differing speeds. A faster wave will eventually catch up with the wave ahead, the two crests coming together to give a 'freak'. It is always possible for two successive waves not to be precisely parallel. Eventually, their crests will meet as another 'freak' peak, which will appear to run along them both.

That waves do not all run at uniform speed and that they do not all lie truly parallel is basically due to the effect of gustiness, of variations in wind speed and direction for long enough to affect the speed and the

Fig. 24. (opposite).
A Characteristic wave properties: 1 *Cycloidal;* 2. *Trochoidal;* 3. *Sinusoidal.*
B Wave height related to wind speed
C Wave length, speed and period.

Wave properties
Waves have the following properties:
Length (L) from crest to crest.
Speed (V).
Period (T), the time between successive crests.
Height (H), trough to crest.
Steepness ($\frac{H}{L}$), ratio of height to length.
Speed, length and period are related, very simply:
$$V = \frac{L}{T} \text{ or } T = \frac{L}{V} \text{ or } L = TV$$

Fig. 25. *Swell waves round a depression. A Occlusion, with S'ly sea ahead of it and N'ly sea behind it. B Confused wave pattern. C SW'ly swell which runs ahead of the warm front. D Confused wave pattern. E NW'ly swell invades the warm sector ahead of the cold front.*

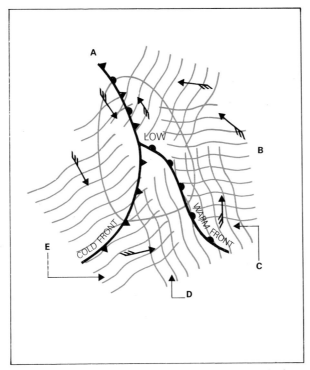

line of the waves beneath it. In seas beneath fast-moving air-streams, especially of unstable air, 'freak' waves are thus quite commonplace.

Wave development around cyclones

A further and most serious complication is that winds of storm force are cyclonic: they move in circles, and as they circulate they are also moving laterally, and then along tracks which are rarely, if ever, on a constant bearing. The trouble is that although strong winds can move in circles, and areas of strong winds

can move across the face of the oceans in curves, the waves they create can only run straight!

So, as cyclones drive on, whether they be tropical or merely vigorous temperate depressions, the waves generated around their four quarters pound on virtually at the speed of the wind which produced them and *in the same original direction*.

Waves can run ahead of fronts, as swell. A south westerly sea generated over a long period and a long fetch in the warm sector of a depression, may have waves 500 ft. (150 m.) long moving at 30 knots. These will run ahead of the warm front (advancing at 20 knots) as a south westerly swell imposing itself over a southerly sea generated by the southerly winds ahead of the warm front. To describe the result as 'confused' is, perhaps, an understatement. Yet into this confusion, a few hours later will come the south westerly waves behind the warm front. Before this has had time to settle as the dominant pattern a further invasion of a north westerly swell, running ahead of the cold front, is inevitable.

Three or even four wave patterns can thus combine to agitate the surface waters and anything floating on them. It becomes impossible to handle a vessel to cope with one pattern of wave movement without exposing it to risks from another.

Hence the greatest risk at sea is the *occlusion*. Here a NW to N 'sea', or 'swell', is invading across a southerly 'sea' ahead of the occlusion, and the frontier line of the 'sea' is ahead of the frontier line of the wind. The swell veers to north westerly, on top of the existing southerly 'sea', before the wind veers.

It is not the waves created by a given wind blowing for a given time and reaching a related length, height and speed which is a sailing hazard; it is several wave trains in the same area at the same time.

Sea and swell waves

Pressure tendency and wind strength

Strong or gale-force winds cause more alarm and despondency, if not actual damage to both boats and reputations, when they occur after a spell of Fine or Fair weather, with moderate to good visibility, favourable wind, and no rain and no fog. For this type of weather the barometer must be relatively high and steady. It therefore cannot rise, at least not much and not suddenly; it can only FALL.

Any fall, however small, must be regarded as the writing on the wall. A fall of the order of 1 mb. in 1 hour is a confirmation that the weight of the air mass is changing and it is getting lighter – the early signs of a Low – somewhere, but where? So watch the wind direction.

Any Low means freshening winds, rougher seas, more and lower cloud, the risk of fog and the inevitable cold front (or occlusion) behind it, bringing fresh north westerly winds, squally showers, but otherwise good visibility.

Two questions become important for the sailor. To what force will the wind increase and how long will this increase take? Other questions follow. Where do I then plan to be and is it safe there? This last, double question can only be answered by the person concerned and the answer will inevitably have to be a compromise between navigation and meteorology.

Timing the onset of strong winds

It is important for the sailor to be aware of the wind conditions produced by a typical depression. He can then evaluate the *actual* conditions in which he is likely to find himself.

He has to bear in mind the following headings:

PRESSURE:	Actual?
	Tendency, i.e. rate of change?
WIND:	Rate of increase, rate of change of direction?
WEATHER:	Rate of deterioration?
CLOUD:	Rate of increase in amount and lowering of cloud base?
VISIBILITY:	Rate of deterioration?
SEA:	Rate of increase of wave height and speed?
SWELL:	Increasing swell suggests that the generating strong-wind area is coming nearer. Swell across the sea foretells a change of wind direction.

Any variation between what is actually happening and what one would normally expect for a typical depression, must then be attributable to one or a combination of the following factors:

1. Change of depth of the centre – is it deeper or shallower?
2. Change of direction of movement – is it crossing to north or to south?
3. Change of speed – is it accelerating or slowing down?
4. Own course and relative speed.

Effect of the coast on wind

When the wind over the open sea lies within 2 points (say $22\frac{1}{2}°$) of the lie of a windward coast-line, even if the land is low-lying there will be an increase of wind strength along that coast of 5 to 10 knots, appreciable up to 10 miles offshore. A coast-line of high cliffs will obviously increase this effect. A coast-line of bays and headlands will create a most uneven pattern of wind strength and direction.

Anticyclonic effect

A pressure gradient, or isobaric spacing which will create a given wind strength for straight isobars will give an additional 5 to 10 knots i.e. another 1 or 2 numbers on the Beaufort Scale, if these isobars are curved around a High Pressure area. Thus, when the barometer is high, winds tend to be stronger than expected.

Gustiness and surface friction

Wind is always gusty even in the free atmosphere above the influence of surface friction, varying in both speed and direction. This quality of 'gustiness' is greatest (up to 50% for speed) in unstable air streams such as Polar Maritime air, and least (as low as 10%) in stable air streams such as Tropical Maritime air.

In *positive gusts* the speed is above the mean (average) speed and in *negative gusts* it is below. In the surface layer of the air – up to about 2,000 ft. (600 m.) – where friction affects the direction of the wind, gustiness in speed produces gustiness in direction. Were there no friction, the wind would always be along the direction of the isobars.

Friction in the surface layers *backs* the wind in the Northern hemisphere and conversely *veers* it in the Southern. Over land the amount is about 15° to 20°, but this obviously depends greatly on local geography. Over the sea the amount is less and 1 compass point ($11\frac{1}{4}°$) is about the average.

In the Northern hemisphere, *positive gusts*, have more energy to overcome the friction and therefore their direction veers i.e. alters to the RIGHT which is the starboard side of a vessel.

Negative gusts, below the mean speed, have less energy and the direction backs, i.e. alters to the LEFT, which is the port side.

In the Southern hemisphere the effect on direction is reversed.

Gustiness and sailing into the wind

Positive gusts are gusts to starboard and will take the vessel that way and with greater force. If the vessel is on the starboard tack, i.e. with the wind coming from the starboard side this will be to her advantage for she will surge forward and her head will move to starboard, *closer* to the mean wind. Similarly on the port tack the ship's head will have to lay off to starboard and off the optimum course into the mean wind.

Negative gusts

Conversely, the *negative gust*, backed and weaker, is less of a disadvantage to the vessel on the starboard tack than it is of advantage to a vessel on the port tack. All of which will explain to some extent the claim so often heard 'that she always goes better on the starboard tack'. It is very possible that this may be a contributory reason for the vessel on the starboard tack being given the right of way and for the 'alter course to starboard' rule.

There was another important factor in the history of the development of the 'Rule of the Road' at sea. Sailing vessels on the great ocean trade-routes of the world normally ran before the wind, an operation for which they were most efficiently designed. They only had to beat into wind when the latter was unfavourable. This became important, however, at the beginning of a voyage, when getting clear into the open sea, or at the end of the voyage, when beating into the destination port.

There are two great concentrations of sea-ports, those of NW Europe (mostly requiring a passage of the

English Channel), and those of the Eastern Seaboard of North America. Both these lie in the Temperate Cyclone Belt and are continuously affected by an endless sequence of depressions in which the wind pattern is predominantly that of: southerly, veering to south westerly, veering to north westerly.

For vessels attempting to beat to the westward down the Channel into the Atlantic, or from the Atlantic into the North American Ports, these changes of wind direction, (the veering), are to starboard, with the final veer to NW, permitting a course of due W. Given sufficient sea room, to this advantage of the starboard tack is added another, namely that the course of this tack takes the vessel away from the depression centre and into an area of better weather and more moderate winds.

The advantage resulting from the wind on the starboard side taking a vessel away from a storm centre, is of the greatest significance, for it was thus the best and safest action to take in the vicinity of tropical cyclones, hurricanes and typhoons, in the *Northern* Hemisphere. It is perhaps not surprising therefore that the modern *The International Rules of the Road* should be in accord with the prevailing weather factors of the Northern hemisphere, and of NW Europe in particular.

Fig. 26 shows two sailing vessels beating up-wind, one on the port tack the other on the starboard.

It becomes clear that positive gusts will give the vessels an added acceleration in speed and the veer of these gusts will take the bows of both vessels off to their right i.e. to Starboard.

The vessel on the port tack just cannot go further to port. Her head would come up directly *into* the wind. She would be 'in irons' with her sails flat-aback and ripe for dismasting. But she *can* lay off to star-board and that right quickly – so the rules require her to do so.

The vessel on the starboard tack has the right of way. Positive gusts will help her to keep further to her right i.e. to starboard and they will help her to keep her speed. She could, of course lay off to port quite easily – into danger – but she doesn't; she maintains her course and speed in safety.

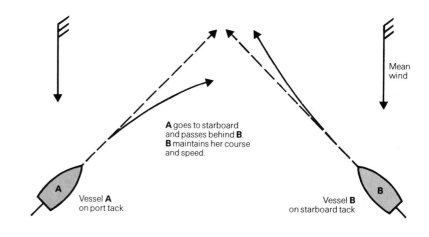

A goes to starboard and passes behind B. B maintains her course and speed.

Mean wind

Vessel A on port tack

Vessel B on starboard tack

Fig. 26. Sailing vessel on starboard tack has right of way.

Fig. 27. Apparent wind –
fore and aft. 1. Sailing dead
into the wind, which is
only possible under power.
2. Running with wind aft.
The vessel's own movement
creates a 'wind' in the
opposite direction, and of the
same speed.

Fig. 28. Reaching: course at
90° to the true wind of 15
knots. The apparent wind
depends on course and
speed in relation to the
true wind. Both vessels are
apparently sailing into the
wind by 18° (1½ points).
In fact, they are making no
ground up-wind.

Fig. 29. Reaching: course
with apparent wind abeam
i.e. at 90° to ship's head.
Wind 15 knots, speed 5
knots. With the apparent
wind abeam, the course is
115° off the true wind. The
vessel is sailing down-wind.

Fig. 30. Sailing into the
wind. Sailing 35° (3 points)
off the apparent wind, the
vessel is in fact sailing 45°
(4 points) off the true
wind. Sailing 'close-hauled'
to within 2 points of the
apparent wind, a ship will be
making good about 3 points
into the true wind.

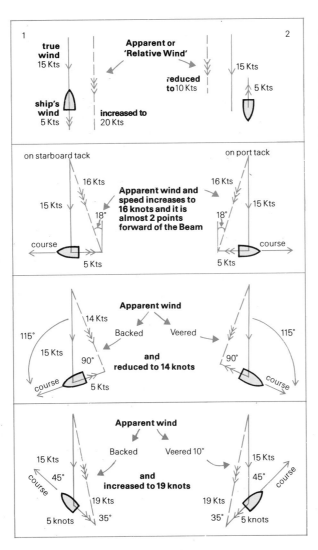

Apparent wind

The moment a vessel begins to move peculiar things seem to happen to the wind. It apparently changes both in speed and direction in ways which are often difficult to understand or evaluate. The apparent reduction in the wind when running with the wind right aft seems perfectly logical to most sailors but what happens when reaching or beating into the wind often seems illogical, or is put down incorrectly to a change in the actual *true* wind.

In fact, the moment a vessel is under way it creates its own wind. A vessel moving at 5 knots in a flat calm creates a 5 knot wind in the opposite direction, and so on. When a wind is blowing this 'ship's wind' must be added vectorially (i.e. by drawing lines to scale) to the true wind, when completion of the triangle called the *relative wind triangle*, gives the resultant *Apparent Wind*.

Figs. 27–30 illustrate how the *apparent wind* depends on the ship's speed:
True wind is shown.
Ship's wind is shown.
Apparent wind is shown.
The diagrams all clearly show how great is the apparent change of direction of the wind. The change in speed is not so marked for slower vessels but is significant for faster craft.

Two things are of particular importance:

1. *Reaching.* Sailing across the true wind at an angle of 90° to it brings the apparent wind forward of the beam and this, quite erroneously, gives the impression that the vessel is sailing into the wind and making ground upwind. See Figs. 28 and 29 which show that when the ship's speed is about one third of the

true wind speed the apparent wind shifts to almost 2 points forward of the beam. When this is actually experienced at sea it is very hard to believe that the vessel is not sailing to windward at all, but only directly across the true wind. A faster craft which can sail at 10 knots across a wind of 15 knots will, on that course, bring the apparent wind almost 4 points forward of the beam and still make no ground to windward.

The apparent wind must therefore be brought to within 3 points of the bow before a vessel can make any ground into the true wind.

2. *Beating into the wind.* Fig. 30 illustrates that a vessel sailing at 5 knots, 4 points, 45°, off a true wind of 15 knots will bring the apparent wind to about 3 points of the bow, 35°. This is approaching the maximum possible into wind, which is something in the region of 2 to 2½ points. Note that the strength of the apparent wind is increased by 1 Beaufort Force.

A faster craft capable of 10 knots in a 15 knot wind will bring the apparent wind for a similar course, to 27° or about 2½ points off the bow, and its strength will be increased by 8 knots or approximately 2 Beaufort Numbers.

In both examples the course being made good is 45° off the true wind and in fact this is the maximum 'into-wind' course that can safely be used in navigation for calculation of course or courses to steer.

Some vessels, with expert crews, can and do sail closer to the true wind than 45° but in normal cruising this high degree of performance cannot be expected. Moreover, allowance must always be made for leeway and this is always a value unique to every boat and one which only the Master can assess from experience.

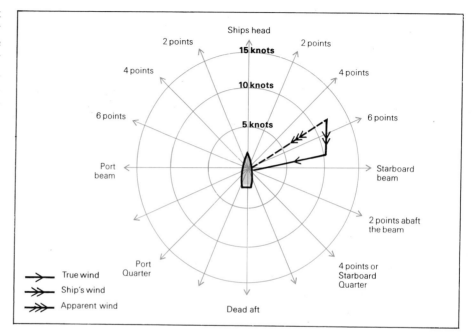

Fig. 31. True wind to apparent wind.
Method: 1. Lay off the apparent wind relative to the ship's head. 2. Add ship's wind, equal to speed but opposite to course. 3. Complete the triangle to give the true wind. 4. True wind direction is: ship's head bearing + starboard relative bearing OR ship's head bearing − port relative bearing.

Apparent wind to true wind

The only way to evaluate the true wind for both its speed and direction is to draw a Relative Velocity Triangle, either mentally or in reality as shown in the diagram above.

If this diagram is used – and there should be a copy placed in every navigation log book, the true wind can be read off and logged in preference to the apparent wind, which is what usually appears in the records.

This true wind is the one which has a relationship to the met. chart isobars and wind scales, and to the recorded records of shore reporting stations.

Fog and visibility scale for ships at sea

Code		Objects not visible at:—	Notes
0	Dense fog	50 yards	*Fog signals obligatory*
1	Thick fog	200 yds	*Speed to be reduced*
2	Fog	400 yds	*Position fixing not*
3	Moderate fog	1000 yds/½ n. mile	*possible except by*
			Radio-Aids or Radar
4	Mist or haze Very poor visibility	1 n. mile (2000 yds)	*Land marks cannot be seen unless vessel is very*
5	Poor visibility	2 n. miles	*close in-shore*
6	Moderate Visibility	5 n. miles	"Coastal" Navigation safe.
7	Good ,,	10 n. miles	
8	Very good ,,	30 n. miles	
9	Excellent ,,	Over 30 n. miles	

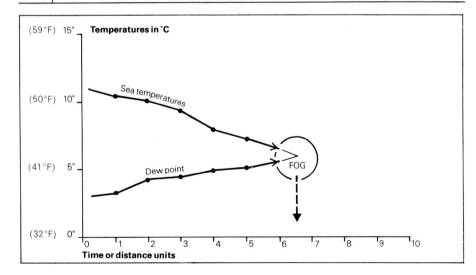

Effect of fog at sea

On land only when visibility falls to below 500 ft. (200 m.) to justify the term *fog*, do certain activities become restricted or potentially dangerous, and these are mainly associated with fast-moving transport on railway and motorway, or with aircraft taking off and landing. Most other activities and sports are unaffected until visibility falls to below 160 ft. (50 m.) — *dense fog.*

At sea these limits are too low.

Fog at sea is officially defined as visibility below ½ nautical mile or 1,000 metres, and in such dangerous conditions vessels are required to reduce speed and to make noises to indicate what type of vessel they are and what they are doing.

Consider what this distance means in terms of speed and time. For two vessels approaching almost head-on, both at 10 knots, it will give them just 3 minutes, for them to sight each other, assess each other, decide what to do and for helm orders materially to alter their courses. At 20 knots the time is reduced to 90 seconds.

Refer back to Fig. 26 on page 77, which shows two sailing vessels on collision course. If they are 1,000 yards apart at sighting, the collision point is 700 yards

Fig. 32. Fog prediction at sea. When sea temperature falls to within a few degrees of the dew-point, fog is likely to form. The time when, or the position where, this will occur can be predicted using hygrometric tables (for the dew-point temperatures) and tables showing average winter and summer sea temperatures.

ahead of each of them or, at 10 knots, about 4 minutes. By letting go the main sheets and putting the rudder hard-a-starboard the vessel on the port tack just has time to veer under the stern of the ship on the starboard tack which is keeping her course and speed.

Fog banks at sea
The formation of fog requires that the air be cooled to its dew-point. If the sea surface temperature is lower than the dew-point, fog will form over large areas, as it does in the warm sectors of temperate depressions. There, Tropical Maritime air with a high dew-point is flowing over sea surfaces which are in general much colder.

However, fog banks can form when the sea temperature is higher than the dew-point, up to a limit of about 9 °F (5 °C). The surface waters are in motion as ocean currents, into which, at high latitudes, great blocks of ice break off the glaciers of Greenland or the vast ice-sheets of Antarctica. As these melt they create vast pools of colder, slightly less salty, surface water within the general circulation. These may be up to 9 °F (5 °C) cooler than the waters which surround them; and they can generate fog banks.

Around the continental shelves the small tidal bulge of the open oceans is translated into tidal streams of moving water. When these streams reach very shallow water, colder water is often brought to the surface by the rising tide over shoals or around islands. Fog banks can form over these colder surface pools.

Inspection of Tidal Stream atlases, in conjunction with the surroundings of Navigational Charts will show where fog banks are most likely to form. Where they will drift to depends on the wind. If it is not always possible to avoid being caught in a forming fog bank, it is usually possible to deduce the most probable location of warmer water, and then set course for it.

Taking temperatures

The reading of temperatures – dry-bulb, wet-bulb and sea – is a worthwhile routine at sea, to be carried out once in each watch and also after every major wind shift i.e. change of air mass.

A high degree of accuracy is not needed, nor indeed is this practicable in small boats. A simple but sturdy instrument is best, of medium size (10 in./25 cm.) for easy reading and handling, with a hanging ring and a fathom of cod-line attached. On the weather side, exposed to the wind but sheltered by your body from the sun, it will give the dry-bulb temperature to 1 °C, or °F if preferred.

A clean piece of cloth thoroughly wetted with drinking water and wrapped around the bulb will immediately convert it to a wet-bulb thermometer, accurate to within 1 °C or 1 °F.

Towed overboard for a minute or so and read quickly when retrieved, it will give a similarly accurate reading for the sea temperature.

Log all three readings. If they reveal that fog is a possibility, prepare for it navigationally and obey the rules of good seamanship. Then attend to the meteorological requirement of reading these temperatures at frequent intervals either of time or of distance – not more than one mile (1,500 m.) apart. This time, not only log the figures but graph them too.

This is an exercise which can obviate a lot of unnecessary speculation and worry before fog occurs, and can greatly aid the process of trying to sail out of it.

Chapter 8 Making your own weather charts

*Sample radio
weather base chart*

GEOSTROPHIC WIND SCALE - SEA LEVEL SURFACE
LAMBERT CONFORMAL CONIC PROJECTION

MAP SCALE = 1:15,000,000
True at 30° and 60° N

82
GEOSTROPHIC WIND SPEED (KNOTS)
* Approximate surface wind speed (knots); about 65% of geostrophic speed

SCALE OF NAUTICAL MILES AT VARIOUS LATITUDES
50 100   200   300   400   500

— — — COASTAL FORECAST AREAS

Basic charting information

ALTHOUGH THE MARINE weather bulletins broadcast by most National Weather Services are primarily for the benefit of merchant and naval ships, there is no reason why such bulletins should not be listened to, understood and used to full advantage by sailors and all amateur meteorologists—or indeed by anyone interested in the weather.

The amateur yachtsman needs also to be an amateur meteorologist and he should certainly learn to use the marine weather services to the full because, unlike his compatriot on land, he cannot always consult the weather charts as published in the press. If he uses a radio-telephone he cannot of course monopolize radio-time; therefore, any information acquired by this means is, of necessity, very terse and to the point.

Recommended schedule

Most experienced yachtsmen realize that weather systems can change dramatically over a short period of time and that for weather charts to be useful they must be prepared regularly—at least every day and more often if weather is likely to be threatening. To prepare an accurate forecast based on the charts, the interval between consecutive charts should be shorter than 24 hours. Differences from day to day in some of our coastal and offshore waters can be so great that it would be extremely difficult to forecast a further 24 hours from a sequence of charts compiled at intervals of 24 hours or more, such as during the hurricane season along the U.S. east and gulf coasts.

From a six-hourly sequence, the problem of forecasting what the next two or three charts will look like becomes very much easier and this period is the basis of professional forecasting techniques.

It is not always feasible for national broadcasting services to transmit such a six-hourly schedule, but many do and if not at six, certainly at twelve-hour intervals. For example, the United States Coast Guard radio at Portsmouth, Virginia (NMN) broadcasts weather synopses and forecasts for various sections of the Atlantic eight separate times per day. These are radio-telephone broadcasts (voice) on single side-band frequencies (SSB).

Radio station, area, and broadcast time	*Frequency and mode*	*Contents*
2–0031 Portsmouth, Va. (NMN) *Area:* (a) North Atlantic north of 03°N, west of 35°W, including Gulf of Mexico and Caribbean Sea. (b) Offshore waters: north of 41°N, west of 60°W. (New England Waters) (c) Offshore waters: 32°N-41°N, west of 65°W. (West Central North Atlantic Waters) (d) Offshore waters: Southwest North Atlantic		
0400	4393.4 (A3J) 6521.8 (A3J) 8760.8 (A3J)	S, F[1]
0530	do.	S, F[2]
1000	do.	S, F[3]
1130, 2330	6521.8 (A3J) 8760.8 (A3J) 13144 (A3J)	S, F[2]
1600, 2200	do.	G, S, F[3]
1730	8760.8 (A3J) 13144 (A3J) 17290 (A3J)	S, F[2]
[1] For areas b and c [2] For area a [3] For areas b, c, and d	S = Weather Synopsis F = Forecast G = Gulfstream Analysis A3J = Single Sideband, Suppressed Carrier	

The radio station at Norfolk, Virginia (NAM) broadcasts the surface analysis weather chart for the North Atlantic twice daily as well as other information. These broadcasts are 'continuous wave' or CW broadcasts and are in Morse code.

Radio station, area, and broadcast time	Frequency (kHz)	Contents
1–0010–1A NORFOLK, VA. (NAM) *Area:* (a) No. Atlantic north of 03°N, west of 35°W, including Caribbean Sea and Gulf of Mexico		
0030, 1230	1213.5 8090 16180 20225[1]	W, A
0630, 1900	do.	W, F
0500, 1700	do.	IB
Broadcast keyed from Londonderry NST [1] 1200–0000 only	W = Warnings A = Surface Analysis Chart IB = International Iceberg Bulletin	

The aim of this chapter is to explain how to listen to such broadcasts, how to draw a surface analysis chart from the voice and the coded broadcasts, and finally how to prepare a forecast of conditions 12 to 18 hours ahead.

Equipment required

1. Radio receiver ⎫ with batteries if
2. Cassette tape-recorder ⎭ required!
3. Recording log-book or pre-printed pad, with pencil attached.
4. Pad of pre-printed charts.
5. Black ball-point pen, and coloured pencils.
6. Soft 2B pencil.
7. Erasers.

1. *Radio receiver*
(a) If an 'alarm-clock' switch can be fitted, time can be saved, and the worry of possibly missing an important broadcast avoided.
(b) A spare ear-phone attachment can be modified to plug into the microphone socket of the tape recorder for direct recording.
(c) The type of radio you will need depends on how far offshore you plan on venturing and to what level of detail you plan for your chart preparation. The United States National Weather Service (NWS) provides continuous weather broadcasts to mariners within listening range of its VHF-FM radio stations in most U.S. coastal locations. These line-of-sight transmissions can be received up to approximately 100 miles offshore depending on the antenna height at the broadcast site. A very inexpensive portable radio with a public service band will be adequate to receive these broadcasts at 162.40 MHz, 162.475 MHz, or 162.55 MHz. To receive the SSB (voice) and CW (Morse code) broadcasts a more sophisticated radio will be needed. The portable multiband marine radios with a 'beat frequency oscillator' will be needed for these broadcasts.

2. *Tape-recorder*
The speed at which some broadcast bulletins are read makes it very difficult for even the most skilled to listen, edit, and write down all of the important information contained in a broadcast. Of course, for the Morse code broadcasts a tape-recorder is mandatory unless you are an accomplished radio operator. The advantages of the play-back cannot be over-emphasized.

3. *Recording log-book*
Some sort of weather log-book should be kept, separate from your ship's log. A three-ring binder or a

spiral notebook is adequate. In small boats, which are always wettest in the type of weather when weather-maps are most needed, it is advisable to have a polyethylene bag of suitable size in which to keep the maps dry, and stowage on a convenient bulkhead which permits the topmost chart to be seen through the protective covering.

To the inside cover of the log-book should be pasted the key to the weather symbols shown on this page, together with any other meteorological information which is likely to be needed.

4. *Weather charts*

Whatever the weather, the recording of weather information should begin at least 12 hours before any cruise is to start and it should be kept going during rests in any port of call.

Blank weather maps should be big enough to cover at least some 20 degrees of latitude, i.e. 20 × 60NM = 1200 N. miles from north to south, and at least the same distance east to west. Because most weather runs from west to east, a map covering a greater span of longitude would therefore cover a longer time-span of moving systems and weather patterns.

A map can be made for any part of the world on any suitable scale from any atlas or reference chart. The sample on page 82 is for the United States, its coastal and offshore waters, and Canada.

The map should only carry the very minimum of printed markings as they may tend to obscure the meteorological information which is to be plotted on the map. Basic printed information should be limited to coastlines, parallels of latitude, meridians (lines of longitude), thin lines to demarcate Forecast Areas, and clear dots to define the positions of reporting stations.

Other information, such as names of areas or countries, should not be necessary, but if absolutely vital they should be indicated as unobtrusively as possible.

Map projection

The projection of this particular map is called 'Lambert Conformal Conic' and is the most widely used 'conic' projection for navigation, although its use is more common among aviators than mariners. A straight line on this projection very nearly approximates a great circle route and the two can be considered identical for many purposes of navigation. Although this projection has been chosen for examples in this chapter, any type map projection can be used for your weather chart preparation, since the radio broadcasts give locations in the form of latitude and longitude. The Mercator projection used most frequently by mariners will work fine in your weather chart preparation if you understand the principles involved.

Distance and scale

The nautical miles scale at various latitudes is given at the bottom center of the chart. Meridians of longitude are drawn at five degree (5°) intervals as are the parallels of latitude. The latter are therefore 300 nautical miles apart. Dots have also been printed on the chart at one degree (1°) intervals as reference points for plotting weather systems.

In the lower left corner of the map is a geostrophic wind scale. This scale can be used to forecast wind speeds based on how far apart isobars are located on your finished map. It will also help you draw the map when only wind speeds are given and not isobar locations.

Fig. 33 Plotting symbols

❟	Drizzle
●	Rain
✳	Snow
△	Hail
▽	Showers
☇	Thunderstorm
⚡	Hurricane
═	Mist
∞	Haze
∿	Smoke haze
☰	Fog
☲	Fog patches
☷	Squall

The marine weather broadcast

As mentioned earlier, there are two types of marine weather broadcasts; voice (SSB) and Morse code (CW). Each one provides similar if not identical warning and forecast information, but only the

---

09 APRIL 1200Z

PART ONE

WESTERN NORTH ATLANTIC GALE WARNING

Gale center over the Gulf of St. Lawrence near 49N 62W 996 MB. 12Z Apr 9 will move northeast about 25 knots. Forecast center near 57N 48W 995 MB by 12Z Apr 10 and near 63N 37W by 00Z Apr 11 as a diffuse low with estimated pressure of 1000 MB. Currently within 450 miles over east semicircle and 650 miles to the southwest winds 25 to 40 knots seas 8 to 12 feet. In 36 hours winds becoming 20 to occasionally 35 knots within 650 miles of the center. Remainder of area. . . . .

PART TWO

FORECAST   NORTH ATLANTIC NORTH OF 32N AND WEST OF 35W

Front extended from gale in Part One thru 45N 56W to 40N 58W and southwestward will move eastward 30 knots . . . but somewhat slower over south portions. Thru the next 36 hours over an area generally north and west of front winds northwest 20 to 30 knots . . . Except as mentioned in Part One . . . winds 20 to 30 knots locally higher in squalls near the frontal zone . . . Seas occasionally 10 feet. High pressure centered 40N 80W 1034 MB will move eastward building over Mid-Atlantic coast of Eastern U.S. by 00Z Apr 10. Forecast high near 40N 73W 1030 MB with ridge extending southeastward thru 35N 73W . . . Remainder of area. . . .

FORECAST   NORTH ATLANTIC SOUTH OF 32N AND WEST OF 35W

Cold front at 12Z Apr 9 through 34N 65W to 32N 77W to 30N 80W will move southeastward and by 12Z Apr 10 extend from 32N 55W to 27N 80W. Winds within 150 miles south of front southwest 10 to 20 knots except stronger in thunderstorms near front . . . Winds north of front in western portion northwest 20 to 30 knots. Remainder of area. . . . .

Fig. 34
Sample plain language voice broadcast (SSB) of offshore weather from the U.S. Coast Guard Radio Station at Portsmouth, VA. The broadcast has been edited to include only information covering the westernmost portion of the forecast area.

Morse code broadcast includes an encoded weather chart. The SSB (voice) broadcast contains a detailed discussion of weather patterns, covering the locations of high and low pressure areas, fronts, ridges, and also including a forecast of their movement, sometimes up to 36 hours ahead. However, it does not include locations of isobars (lines of equal pressure); therefore, in drawing your weather chart, some subjective analysis and interpretation are needed and this skill usually takes some time for the amateur to develop.

The CW (Morse code) weather chart messages are broadcast in a coded format (using numbers) which gives the latitude and longitude of points through which each line (front, isobar) passes. Careful plotting and line-drawing should give an exact reproduction of the original map.

Recording a plain language (voice) broadcast

A weather chart prepared from the voice broadcast is probably what most yachtsmen will initially be prepared to do, because the voice broadcasts are more familiar to them. These are the broadcasts that most serious offshore sailors listen to while at sea. In the next column (Fig. 34) is reproduced an edited marine weather broadcast from the U.S. Coast Guard radio station at Portsmouth, Virginia(NMN). These broadcasts are separated into geographical areas as shown on the sample base chart (see page 82) by dashed lines. In this example, the weather map we prepare will cover the area of the 'West Central North Atlantic'. Weather and forecast information covering the remainder of the forecast area has been edited out. The information recorded here was extracted from the plain language broadcast at 1600Z and 2330Z on April 9, 1977.

How the forecast is presented
Part 1. Summary of *Gale and Storm Warnings* in force.
Part 2. General Synopsis and Forecast.
The pressure values, position, estimated direction and speed of movement of the major systems are listed. This is the basic information upon which the National Weather Service staff have based their Sea-Area forecasts of wind (direction and speed), weather, visibility, and sea state for the next 24 hours. Forecasts are always given by the areas listed such as: North Atlantic north of 03°N, west of 35°W, including the Gulf of Mexico, and the Caribbean Sea. A listing of weather broadcast stations can be found in the *Worldwide Marine Weather Broadcasts*, a joint publication of the U.S. Department of Commerce, National Oceanic and Atmospheric Administration and the U.S. Department of the Navy, Naval Weather Service Command. This publication is available through the Superintendent of Documents, U.S. Government Printing Office, Washington, D.C. 20402. In Canada similar information should be available from The Meteorological Service of Canada, 315 Bloor Street, Toronto 5.

Information to be included on the weather chart

1. *Date-time* at the top, using the form of two digits for the day of the month and four digits for the hour. Use the letter Z to denote Greenwich Mean Time (GMT) as GMT is universally used in marine broadcasts. The broadcast will give the valid times, in GMT, of the warnings and forecasts.

091800Z means   *date*: the 9th of the month
                        time: 1800 GMT

2. *The Synopsis.* Plot this information with one of the coloured pencils or with solid lines in pen or pencil (forecasts with dashed lines). First, locate and plot the positions of pressure systems and fronts (see page 88), their central pressures and any other information provided as a general description of the area. Pencil in the radius of wind velocities such as: 'currently within 450 miles over the eastern semi-circle and 650 miles to the south-west (of the gale) winds 25 to 40 knots.' After penciling in these areas,

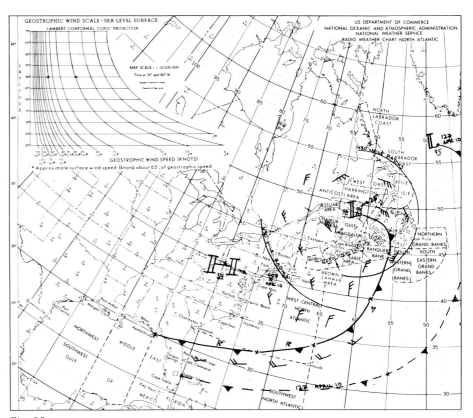

Fig. 35
How your 'voice' weather chart should look after the synopsis information has been plotted. All given information should be on this map.

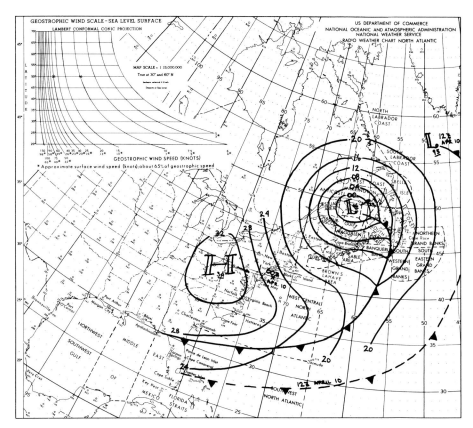

Fig. 36
Completed 'voice'
weather chart
based on the actual
information broad-
cast, as well as
your subjective
analysis of these
data.

draw in the wind barbs as shown on Fig. 35. Remember that wind flows counterclockwise around low pressure areas such as this gale and clockwise around high pressure areas. Pencil in other wind barbs where wind speeds and directions are given in the synopsis. At this point there will be blank areas where you have not received any information. This is where you must interpolate using your basic knowledge of wind flow. Air flows in smooth lines from one area to another. The map on page 87 is representative of what your weather chart might look like at this point in the analysis.

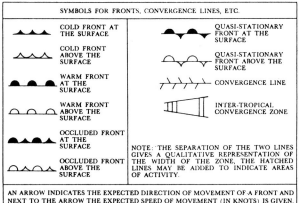

Fig. 37. Symbols used in the plotting of pressure systems and fronts.

3. *Isobars.* By inspecting the direction of your plotted wind arrows and more importantly the velocities given in the synopsis, begin to sketch in the main isobars. This is where a soft 2B pencil and eraser will be needed. Remember that if troughs or ridges are mentioned in the broadcast, include them in your map. A description of each of these features as well

as other features can be found in the last table (page 93) of this chapter. Draw in isobars at 8 mb intervals at first using 1024:1016:1008 etc., then put in the intermediate 4 mb isobars. Keep them flowing, evenly spaced, parallel, and representative of the wind speeds given in the broadcast. The geostrophic wind scale shown in the upper left hand corner of the sample chart shows how far apart isobars should be located at various latitudes to give the wind speeds indicated in the forecast and synopsis. In this example note that the central pressure of the gale and the high are 996 mb and 1034 mb respectively. Therefore at 4 mb intervals you will need to draw an isobar at: 1000, 1004, 1008, 1012, 1016, 1020, 1024, 1028, and 1032 mb or a total of 9 isobars. If you start with 8 mb intervals it is much easier to split the difference between the two pressure areas.

4. *Pressure systems.* Lastly draw in the forecast positions of the pressure systems and fronts with another coloured pencil or dashed lines. The completed chart should look like the one on page 88.

You will have another forecast in 6 or 12 hours time and the synoptic situation in this one should describe what you have just drawn. If it is different, correct your chart – this time boldly – and be ready to use it as the springboard for the next one ahead.

This process should take around 30 minutes and the chart now carries the latest information. Anything more must be added in pencil, and speed and accuracy will come with practice. While no charts drawn by different people from this same basic information will be exactly identical, they should all be more or less alike and any forecasts made from them should agree within reasonably narrow limits. All should reveal the possibility, or probability, of unusual weather conditions or of weather hazards.

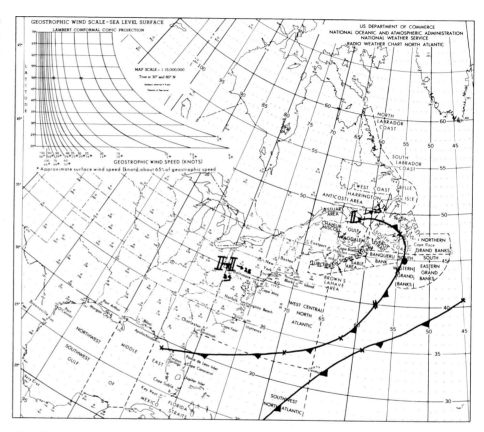

*Fig. 38
A portion of a Morse code weather map that only includes positions of major pressure systems and the frontal group.*

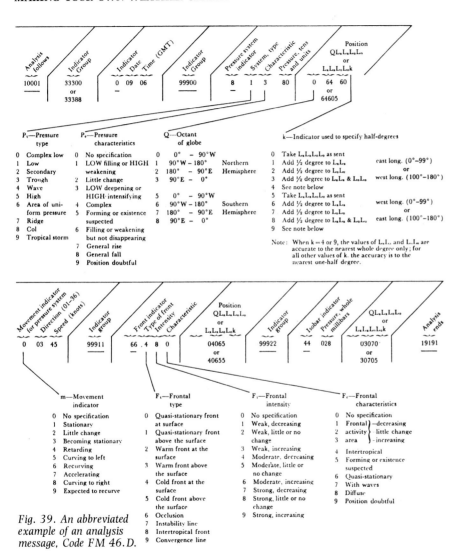

Fig. 39. An abbreviated example of an analysis message, Code FM 46.D.

Recording a Morse code marine weather broadcast

Radio-telegraph broadcasting in Morse code has been the mainstay of marine communications for decades. It has only been in relatively recent years, with the development of solid state electronics, that SSB voice broadcasts have become available to the budget-conscious small boat owner. It is now possible to use both voice and Morse code broadcasts with relative ease. One can record and plot the voice weather synopsis and forecast information, as was done in the last section, and then record and plot the Morse code surface analysis weather chart on a radio-telegraph frequency. In this way one doesn't need to know the entire Morse code vocabulary but only needs to learn the digits from 0 through 9. Since the weather charts are encoded in the format shown on the opposite page and only numbers are used, the decoding is relatively simple.

International Code FM 46.D (Fig. 39)

The weather chart message usually follows the weather forecast and warning information and is broadcast in the International Analysis Code, IAC Fleet, FM 46.D. The map broadcast consists of a series of five digit groups as per the format illustrated in Fig. 39. The 10001 group indicates the beginning of the weather chart broadcast and the 19191 group indicates the end of the broadcast.

Indicator groups and figures shown with heavy underscoring always appear in coded analysis messages. Additional latitude and longitude groups follow in the messages to indicate the points through which fronts and isobars should be drawn. In analysis messages, isobars are usually given for 8-millibar intervals. Intermediate isobars can then be sketched.

The Radio broadcast

In the U.S. Government publication *Worldwide Marine Radio Broadcasts*, there are listed all radio stations that broadcast weather charts. A sample listing was shown on page 84. Under the contents column the letter 'A' designates the broadcast of a surface analysis chart. As you can see, this particular station at Norfolk, Virginia, broadcasts a chart twice daily. These broadcasts also include the weather forecast text that is broadcast on the voice stations but you do not need to copy this information as you have already done it with your voice broadcast recording. In addition, you would need to learn the remainder of the Morse code vocabulary.

As with the voice broadcasts, it is best to monitor the CW broadcast with headphones as you are recording, so that you can adjust volume and squelch as needed. The broadcast itself always follows the same format and after you have decoded several maps you will begin to recognize the various indicators with relative ease. Below can be found a sample message for April 9, 1977 at 1200Z.

As can be seen, the Indicator Groups are underlined, while the sections in each Group are separated by an oblique stroke. Decoding (i.e. translating into plain language) and plotting are simultaneous processes enabling pressure systems, isobars and fronts to be accurately marked on the chart.

Decoding the weather chart

The first thing to do is to place the blank chart in front of you along with the coded message and the Code FM 46.D format. As always the first five digit group, 10001, indicates the beginning of the chart message. The second group, 33388, (which note should always be the same for the United States) indicates whether the latitudes and longitudes will be given in whole degrees or half degrees; they are almost always whole degrees. The date-time group indicates the 09 day at 1200 GMT. The next indicator group 99900 begins the first of three main sections of the message. These are:

99900 = Pressure System Indicator
99911 = Frontal System Indicator
99922 = Isobar System Indicator

Pressure systems and fronts

Step 1. Plot the fix points for pressure centers and enter the central pressure on the chart. Also indicate with an arrow the direction of movement and speed.

Step 2. Plot fix points for each front and draw in a smoothed line of best fit for each front. Mark each front with the conventional notations and hash marks (see page 88) indicating the transition from one type of front to another. The map on page 89 illustrates the analysis to this point.

10001	33388	/ 00912	99900	/ 85134	04080	09025	/ 81296	04962
04525	99911	/ 66672	04962	04855	04455	04058	/ 66442	04058
03662	03369	02984	/ 66411	04145	03559	02570	99922	/ 44000
05162	05058	04861	04863	05064	/ 44008	05262	05157	04955
04757	04664	04966	05166	/ 44016	05464	05255	04852	04455
04465	04769	05368	/ 44020	02963	03156	04552	04951	05456
05568	05171	04367	04063	03563	03068	02674	02382	/ 44024
05073	04269	03967	03669	02986	/ 44028	04575	04172	03773
03580	/ 44032	04280	04075	03778	03885	19191		

Fig. 40. A sample Morse code message of the weather chart, transmitted via the marine broadcast. Decoding of this message will yield a complete surface analysis chart.

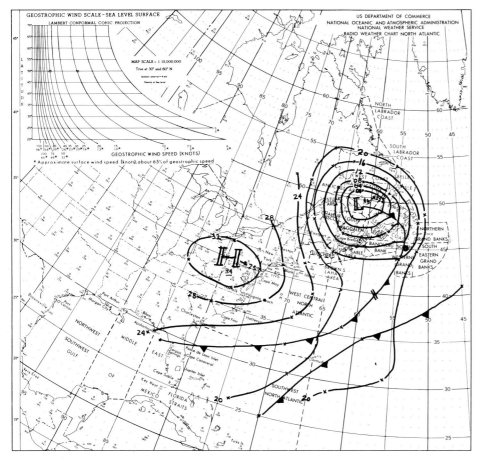

Fig. 41
Completed Morse
code prepared
weather chart for
April 9, 1977 at
1200 GMT.

Isobars

Following the isobar system indicator, 99922, and the first isobar group indicator, 44000, can be found the first fix point for that isobar; 05162. In other words the first fix point for the 1000 mb isobar is located at 51° North Latitude and 62° West Longitude. Plot this point and enter its pressure value nearby. Proceed by plotting the rest of that isobar's fix points, then draw (lightly) in pencil a smoothed curve to fit the fix points. Go on to the next isobar, etc.

Remember as you work that isobars are smooth curves except at those points where they cross fronts. Here they should be drawn with an angle in such a way that the acute angle is facing towards low pressure (Fig. 42).

Also isobars should not cross each other. If yours do there is an error in coding, transmission or plotting of the analysis. Check your plotting first. If you do not find a plotting error try to adjust the faulty point to best reconcile the isobar in question with other nearby isobars. The most common errors in both encoding and decoding are to reverse the latitude and longitude numbers or to read a value either 5 or 10 degrees off position.

Where isobars cross fronts the analysis will usually give a fix point or intersection of isobar and front. This point frequently needs minor adjusting since positions are given to the nearest whole degree. Adjustment may be made to either the isobar or the front so as to produce the smoothest set of lines.

Isobar intervals of 4 or 8 mb are generally used in the broadcast analysis. In weak pressure fields an intermediate isobar at 2 mb intervals may be coded to more closely define the wind field. In such a case be careful to label the 2 mb interval isobar correctly

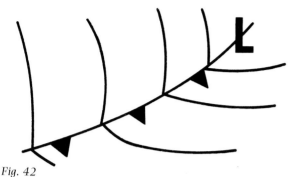

Fig. 42

and draw it as a dashed line to distinguish it from the regular 4 mb interval.

After all the isobars have been drawn in and smoothed they can be gone over more heavily. The map on page 92 is representative of a Morse code produced weather chart. Notice how it differs from the map (page 88) prepared using the voice broadcast alone.

Although the two charts are slightly different the basic information needed for forecasting the weather can be found on both charts. Any two people given the same information will probably draw charts that will differ in several areas. The important thing to note is that each broadcast format, both voice and Morse code, carries enough information to produce a weather chart that is useful for offshore boating. The chart combined with the forecast of movements of the various systems should provide you with sufficient information to tackle most any cruise ranging from intercoastal to transoceanic. All serious yachtsmen should be aware of the various weather broadcast services available to them and in addition, know what to do with this information. Hopefully this chapter has provided some of that information.

Terms used in weather bulletins
Cyclones or *Depressions* or *Lows* are:—
SHALLOW Centre pressure not less than 1000 mbs.
DEEP Centre pressure 960 mbs.
VERY DEEP Centre pressure 930 mbs. or less.
DEEPENING Central pressure decreasing.
FILLING Central pressure increasing.
VIGOROUS Characterised by strong winds.
COMPLEX Has two or more centres within the main cyclonic circulation.

Troughs are where the Isobars are V-shaped. There is usually a change of air-mass, i.e. a Front, along a Trough and always a marked change in the wind direction. They usually join near the centre of young depressions but trail behind old filling ones.
Ridge: An area of higher pressure between depressions. Here the air is drier (less cloud), often colder and the weather is Fair to Fine.
Col: An area of lower pressure between cells of high pressure or of higher pressure between two lows.

Barometer tendency
Rate of change in 3 hours:
STEADY Less than 0.1 mb.
RISING OR FALLING SLOWLY 0.1 mb. to 1.5 mbs.
RISING OR FALLING 1.6 mbs. to 3.5 mbs. i.e. about 1 mb./hour, the warning of a depression.
RISING OR FALLING QUICKLY 3.6 mbs. to 6.0 mbs. i.e. Approaching 2 mb/hour. warning of force 6.
RISING OR FALLING VERY RAPIDLY Over 6.0 mbs. i.e. Approaching 3 mb./hour, warning of Gale Force 8

Fronts
Warm Front: This is the advancing boundary of Tropical Maritime air. It is characterised by falling barometer, lowering stratus cloud with rain, freshening and backing winds S or even SE'ly. There is always a risk of fog at sea if the sea surface temperature is low. Visibility deteriorates.
Warm sector: This is the area of Tropical Maritime air behind the warm front and ahead of the cold front. The isobars are usually almost straight, equidistant and parallel so that over the whole sector the wind is generally uniform in both direction and strength.
Cold front: The advancing boundary of the cold Polar air, characterised by a marked wind veer, squally showers, often as line-squalls, confused seas where NW wind and sea is invading SW wind and sea, and a dramatic improvement in visibility. There is no general fall in pressure ahead of it to give warning of its approach. When the cold front catches up with the warm front the cold front weather is then embedded within the warm front weather — the worst of both worlds coming together, and the wind veers very abruptly from S to NW. This is particularly hazardous.
Cold occlusion: When the Polar air behind a depression is very cold, very heavy and the wind speed is high, it can dig beneath the less cold returning Polar Air ahead of the depression. This is the usual form in Mid-Atlantic.
Warm occlusion: When the returning Polar air ahead of the depression is colder and heavier than the Polar Maritime air behind it, the latter will lift over the former. This occurs when the depression approaches land and the returning Polar air is 'Continental' rather than 'Maritime'.

Index

Credits

The author and publishers would like to express their grateful thanks to the following sources for advice, material, information and other assistance willingly and courteously afforded during the preparation of this book:

Director General of the Meteorological Office, particularly for permission to reproduce photographs, charts and diagrams.

Royal Meteorological Society and Royal Yachting Association, particularly for permission to reproduce from their basic met-maps.

BP Educational Service, whom the author wishes especially to thank for their help with, and permission to reproduce, a selection of their excellent Cloud colour photographs/transparencies which were taken by Mr. R. K. Pilsbury. BP Educational Service is a registered name of BP Trading Limited, a member of the BP Group of Companies.

United States Navy Department for permission to reproduce the photograph of 'swell waves' on page 71.

Illustrations: Thanks are also due to the following for supplying and/or granting permission to reproduce the illustrations used in the book. Artwork for drawings, diagrams and maps was prepared by Arka Graphics from references indicated. Credits are given spread by spread.

Cover: Spectrum Colour Library; Barry Walker; Arka Graphics

1 Philip Clucas

2–3 Philip Clucas

4–5 Philip Clucas

6–7 Meteorological Office/Crown Copyright Reserved

8–9 Casella (barograph trace)

12–13 BP Educational Service/ R.K. Pilsbury

16–17 Syndication International

20–21 BP Educational Service/ R.K. Pilsbury

24–25 BP Educational Service/ R. K. Pilsbury

28–29 BP Educational Service/ R. K. Pilsbury

40–41 Syndication International

44–45 Philip Clucas

52–53 Syndication International

64–65 Spectrum Colour Library

66–67 Meteorological Office (4); Spectrum Colour Library (2)

68–69 Meteorological Office (4); Spectrum Colour Library (2)

70–71 Meteorological Office (3); U.S. Navy Dept.; Meteorological Office/Syndication International; Meteorological Office/Crown Copyright Reserved

82–93 Maps based on National Weather Service Radio Weather Charts and prepared by Robert Mairs. Figs. 37 and and 39 reproduced from *World Wide Marine Weather Broadcasts* (U.S. Government publication).